Technical and Business Writing for Working Professionals

Technical and Business Writing for Working Professionals

Ray E. Hardesty

Copyright © 2010 by Ray E. Hardesty.

Library of Congress Control Number:		2010917175
ISBN:	Hardcover	978-1-4568-1939-2
	Softcover	978-1-4568-1938-5
	Ebook	978-1-4568-1940-8

All rights reserved. No part of this book may be reproduced or transmitted in any form or by any means, electronic or mechanical, including photocopying, recording, or by any information storage and retrieval system, without permission in writing from the copyright owner.

This book was printed in the United States of America.

To order additional copies of this book, contact:
Xlibris Corporation
1-888-795-4274
www.Xlibris.com
Orders@Xlibris.com
90446

Contents

INTRODUCTION 9
   WHAT YOU WILL LEARN 10
      Part 1: Basics of English 11
      Part 2: Technical Writing 11
      Part 3: Business Writing 12
   WRITING AND COMMUNICATING 12

PART 1: BASICS OF ENGLISH

   PARTS OF SPEECH 15
      Nouns 15
   PARTS OF SPEECH (CONT'D) 21
      Pronouns 21
   PARTS OF SPEECH (CONT'D) 28
      Verbs 28
   PARTS OF SPEECH (CONT'D) 38
      Adjectives 38
   PARTS OF SPEECH (CONT'D) 44
      Adverbs 44
   PARTS OF SPEECH (CONT'D) 50
      Prepositions 50
   PARTS OF SPEECH (CONT'D) 55
      Conjunctions 55
   SENTENCE STRUCTURE 62
      Introduction 62
      Types of Sentences 62
      Phrases 65
      Things to Remember About Clauses and Phrases 67
      Put Main Idea in Main Clause 67
      Use Simple Sentence Structure 67

Use Parallel Construction 68
Avoid Sentence Fragments 70

PART 2: TECHNICAL WRITING

INTRODUCTION 77
THE FIRST TWO LAWS OF GOOD TECHNICAL WRITING 79
SHORT FORMS 80
   First Use..80
   Initial Caps 80
   Articles With Acronyms 80
   Units of Measure 81
   Degree Symbol...81
   Plurals..81
   English and Metric Units 81
   Period With Units of Measure...82
   Standards...82
GRAMMAR AND PUNCTUATION 84
   Syntax and Usage 84
GRAMMAR AND PUNCTUATION (CONT'D)...88
   Capitalization 88
GRAMMAR AND PUNCTUATION (CONT'D)...89
   Punctuation in Technical Writing 89
NOMENCLATURE 94
   Clarity and Comprehension...94
   Capitalization 94
   Controls and Indicators 94
   Switches 94
   Pots and Selector Switches...95
   Lights and LEDs..95
   Readouts and Meters 95
   Verify That..95
TECHNICAL WRITING STYLE 97
   Active Voice 97
   Clarity and Conciseness...97
   Flammable/Inflammable 98
   Introductory Clause...98
   "It" as a Subject...98
   One Action per Procedural Step 98

  Procedural Steps..98
  Specific Verbs...99
  Starting Sentences...99
  State Actions Specifically..99
  Telegraphic Style...99
 Format...102
  Standard Formats..102
  Military Decimal Format..102
  Alphanumeric Format..103
  Typefaces and Fonts...104
  Line Length and Margins..104
  Table Format..105
  Figure Format...105
  Warnings, Cautions, and Notes...106
 The Writing Process..109
  The Audience...109
  Writing as a Process..111
  Organizing...111
  Creating...112
  Reading..112
  Revising...112

PART 3: BUSINESS WRITING

 Identifying the Audience..117
  Greetings . . . or, "Hello-o-o-o?".......................................117
  The Checklist Method..118
  Why Bother?..118
 Business Writing "Tone"...122
  Defining Tone..122
  Formal Tone..122
  Informal Tone..124
  Expressing Enthusiasm..125
  Conveying Negative Information......................................126
 The Importance of Brevity..131
 E-Mail Correspondence and Interoffice Memorandums........134
  E-Mail..134
  Interoffice Memorandums...136

The Formal Business Letter 146
    Tone 146
    Letter Format 146
    Letter Content 150
The Internal Report 156
    Introduction 156
    Report Style 156
    Report Structure 157
The External Proposal 166
    Introduction 166
    Proposal Content 167
    Boiler Plate Material 171
    Proposal Writing as "Writing" 172
    Editing and Graphic Design 173
Presentations 176
    The Presentation as a Piece of Writing 176
    Characteristics of Presentations 177
    A Presentation as Live Theatre 179
    Presentation Delivery 180
    Using Humor 183

APPENDIX: WRITING TUNE-UPS

Writing Tune-up No. 1: Using Hyphens 189
Writing Tune-up No. 2: The Ubiquitous Comma 191
Writing Tune-up No. 3: Sentence Structure Problems 194
Writing Tune-up No. 4: Relative Clauses 197
Writing Tune-up No. 5: Nonsexist Language Solutions 199
Writing Tune-up No. 6: Verb Tense When Referring to
    Past Research 201

INDEX 203

Introduction

This book was written for two audiences: working professionals of all types and college or university students preparing to enter the worlds of business, manufacturing, technical professions, or research. My primary goal was to serve the audience of working professionals who may need a concise refresher course on the use of the English language in general and the use of language in the professional world in particular. I also hope that it can serve as an adjunct text for college courses in technical or business writing offered to students preparing for professional careers.

In the past decade a change has taken place in the business world that has resulted in ever-increasing writing responsibilities being placed on the shoulders of professionals for whom writing is not their primary specialty. Often, these professional workers feel uneasy and ill-equipped to carry out these writing responsibilities. Many of these professionals are numbers people, more comfortable designing, measuring, and specifying (which is not the same as writing a specification document). Yet, the writing responsibilities are not going to go away. Rather, they are going to increase.

What happened was this. In the 1990s, when desktop publishing programs such as PageMaker became popular, many companies decided that they no longer needed their word processing personnel. The reasoning was that technical writers, armed with the new software, could not only write but also format and publish the company's documents. Today, many technical writers are also asked to spend much of their time creating technical illustrations, complex templates for online distribution of documents, and online help content. Finally, throughout the '90s many companies, facing cost constraints, downsized their technical writing departments or eliminated them altogether, choosing instead to use contract writers when necessary.

This trend may have been inevitable, given fierce competition in the global economy, but the fallout is that professional staff members receive much less support from technical writing staffs than in the past. The writers either do not have the time or have been given, formally or informally, new job descriptions with new priorities. This means that the task of writing is falling more and more on specialists in the engineering department, the human resources department, etc.

Working professionals probably deserve better than this, but in today's environment the situation is not likely to change. The days when technical professionals, for example, could expect a technical writer to work alongside them during the design and manufacturing process are almost gone. The writing responsibilities are now everywhere being moved into the orbit of the engineers, programmers, and other company professionals.

It is my hope that this book will help those professionals who find themselves in this situation. In trying to alleviate the stress many professionals feel when confronted with writing responsibilities, I have created a rather rigorous "course" of study. Although the book may be seen (and I hope it will be seen) as a reference manual to be consulted on an ongoing basis, I intended that the reader should read and study the book in its entirety. To get the most out of the book, the reader should begin at the beginning and read to the end, using the Memory Solidifiers at the ends of the chapters to fix the basic principles learned into memory. It is only when this body of information has been assimilated that one can truly say that one is a good technical and business communicator.

What You Will Learn

This is not a comprehensive course in either technical or business writing. In the area of technical writing, for example, there is no discussion of the various types of technical documents, such as specifications, procedures, and manuals, that must be prepared in the normal course of business and manufacturing. I have assumed that working technical professionals will already be familiar with these document types. It also seemed to me, based on more than twenty years in technical and business communication, that technical document content and format vary considerably across industries and organizations. Thus, for the audience of working professionals I did not feel it would be beneficial to discuss generic document types, preferring rather to assume that technical professionals will learn the details of preparing these documents at their places of employment. For students, of course, a good, comprehensive technical writing text will be necessary to introduce them to these document types. Finally, there is no discussion of vocabulary, irregular verb forms (although verb forms are discussed), and the aesthetics of prose "style." It is left to the reader to assess his or her knowledge and abilities in these areas and to determine if additional work is necessary.

Part 1: Basics of English

In the first half of the book you will become re-acquainted with the mechanics of the English language, including parts of speech and sentence structure.

You may wonder why so much time has been allotted to the basics of language. The most important reason is that rational thinking is based on language. Of course, there are many different "languages," not all of which are strictly verbal. Mathematics is a language. A mathematician or technical professional can write an equation on a piece of paper, hand it to colleague, and information can be transferred without any verbal exchange between the two. Similarly, music and art of all kinds can be considered languages of sorts.

In the world of commerce, however, we must use primarily verbal language to convey information of all kinds to audiences of many types, and this leads us to the second reason we must know the mechanics of language. In order to think and communicate in any language we need to be able to use the tools of that language. In fact, the better we know and can use the available language tools, the better we think and the better we communicate. In short, a language is not just a bunch of arbitrary rules. It is nothing less than the organizational structure we use to understand the world and express that understanding to others.

Part 2: Technical Writing

In this section you will learn about the following topics:

    Short forms (abbreviations and acronyms)
    Grammar and punctuation in technical writing
    Nomenclature

Technical writing style
Format
The writing process.

Part 3: Business Writing

In this section you will learn about:

Identifying the audience
Business writing "tone"
The importance of brevity
E-mail correspondence and interoffice memos
The formal business letter
The internal report
The external proposal
Presentations.

Writing and Communicating

As important as it is, the ability to write good sentences and punctuate them well is not all that is needed to become a top-flight business and technical communicator. Skills such as being able to organize material and analyze the audience are equally indispensable. The most important thing to remember is that communication does not take place until the target audience comprehends and can use our information.

Part 1

Basics of English

Parts of Speech

Nouns

Definition: Nouns identify a person, place, thing, idea or concept, activity, or quality.

Types of Nouns

There are two broad classes of nouns: proper nouns and common nouns. Proper nouns name specific persons, places, things, etc. They are almost always capitalized. (Note: in the non-technical world, it is normally fairly obvious what is or is not a proper noun requiring capitalization; in technical writing, however, this issue becomes more complex. This will be discussed later under Capitalization.)

Examples of proper nouns: Albert Einstein, the Chrysler Building, the Magna Carta, Iowa, Mt. Everest.

All other types of nouns are called common nouns. There are four general classes of common nouns, as follows:

- *Concrete nouns*, which name tangible objects (e.g., tree, carousel, automobile, telephone).
- *Mass nouns*, which name materials in general rather than specific objects (e.g., steel, paint, clothing, wheat).
- *Collective nouns*, which name groups of things regarded as single units (e.g., committee, quartet, team, personnel).
- *Abstract nouns*, which name intangible things such as ideas, activities, or qualities (e.g., optimism, exercise, truth, attention).

Note that many abstract nouns are also known as *gerunds*, or *verbal nouns*. They consist of the "ing" form of a verb that is used as a noun (e.g., manufacturing, fishing, waiting, selling).

Linguistic Functions of Nouns

Nouns are used in the following ways in English:

- *Subject of a sentence:* "The engineer designed a new actuator." (The noun "engineer" identifies who or what carries out the action indicated by the verb.) "Electro-magnetism was the operating force involved." (The noun "electro-magnetism" is the subject of a form of the verb "to be.")
- *Object of a verb:* "He designed the actuator." (Direct object—the noun "actuator" answers the question "what?" or "whom?" about the verb and its subject.) "The company gave our department a new copier." (Indirect object—usually a person or persons, such as "our department"; answers the questions "to whom or what?" or "for whom or what?")
- *Object of a preposition:* "The new accounting system will be active *within* a month." "*At* that time, everyone will benefit *from* the results *of* our extensive pre-planning." (Objects of prepositions are nouns that are related to other sentence elements in some way by prepositions; in the first example, the preposition "within" relates the object "month" to the first part of the sentence.)
- *Objective complement:* "Her superiors recognize her as a leader." (The noun "leader" completes the meaning of the verb's object—"her".)
- *Subjective complement:* "Her brother is also an engineer." (The noun "engineer," which follows a linking verb such as a form of "to be," completes the meaning of the sentence's subject—"brother.")
- *Appositive:* "Harry, the company's treasurer, has decided to retire." (The noun "treasurer" follows and amplifies the meaning of another noun or noun phrase—in this case "Harry." The appositive serves the same grammatical function as the noun it amplifies; here it is the subject of the sentence.)

Noun Usage

Forming Plurals

Most nouns form the plural by adding *s* (e.g., penguins). Nouns ending in s, z, x, ch, and sh form the plural by adding *es* (e.g., churches).

Nouns ending in a consonant followed by a *y* form the plural by changing the *y* to *ies* (e.g., deliveries).

Some nouns ending in *o* add *s* to form the plural, others add *es* (e.g., dynamos, tomatoes).

Some nouns ending in *f* or *fe* add *s* to form the plural, others change the *f* or *fe* to *ves* (e.g., riff/riffs, calf/calves).

Some nouns require internal change to form the plural (e.g., child, children).

Some nouns do not change in the plural (e.g., antelope).

Compound nouns (i.e., nouns made up of more than one word) that are not run together to make one word form the plural in the principal word (e.g., holes-in-one, post offices).

Compound nouns written as one word form the plural by adding *s* (e.g., drugstores).

Possessive Noun Forms

The possessive (or genitive) form (or "case") for most nouns is formed by adding *'s* for living things and creating an "of" prepositional phrase for inanimate objects. Examples:

> The tenor's performance was moving. He brought out the lyrical beauty of the aria.

Note, however, that both constructions are acceptable in most situations.

Plural nouns ending in *s* require only an apostrophe at the end (e.g., "the *managers'* policy statements . . .").

Plural nouns not ending in *s* must have both an apostrophe and an *s* to form the possessive (e.g., "the *linemen's* excellent tackling . . .").

With group words and compound nouns, add *'s* to the last word (e.g., "Chairman of the *Board's* limousine . . .").

Compound Nouns

Compound nouns may be formed by two or more unconnected words, by words run together to form a new word, or by words connected by hyphens. All three forms fulfill the same grammatical functions.

Compound nouns are probably the area where the English language is changing most rapidly, due to the continuing introduction of many new technical and computer terms. This has led to much variation in usage among organizations and even individuals. We will discuss this issue later when we talk about punctuation, but for now, here are some helpful hints:

- For compound nouns in general usage, consult a good dictionary, such as Webster's New Collegiate Dictionary.
- For technical terms, consult specialized dictionaries or organizational style guides.
- When in doubt, do not run words together (i.e., separate the words with spaces or hyphens). Run-together words are almost always harder to read.

Memory Solidifier

Instructions: Under each of the headings below, paraphrase (i.e., write in your own words) the main points under that heading in the material you have just studied. If necessary, return to the previous material to refresh your memory.

Give the definition of a noun:

-

What are the two broad classes of nouns?

-
-

List the four classes of common nouns:

-
-
-
-

What do we call an abstract noun with an "ing" ending?

-

What are the linguistic functions of nouns?

-
-
-
-
-
-

List any points on forming plurals of nouns that you think you may have trouble remembering:

List any points on using possessive forms of nouns that you think you may have trouble remembering:

List any points on using compound nouns that you think you may have trouble remembering:

Parts of Speech (cont'd)

Pronouns

Definition: Pronouns are words that take the place of nouns. They can serve all of the functions served by nouns.

Types of Pronouns

Pronouns fall into numerous categories. Following is a discussion of each pronoun type.

Personal Pronouns
There are three groups of personal pronouns:

- The following personal pronouns refer to the person(s) or thing(s) performing an action: *I, you, he, she, it, we, you (plural), they.*
- These personal pronouns indicate possession: *my, mine, your, yours, his, her, hers, its, our, ours, their.*
- These personal pronouns indicate that a person or thing is the object of an action (i.e., object of a verb) or of a preposition: *me, you, him, her, it, us, you (plural), them.*

Demonstrative Pronouns
The demonstrative pronouns (*this, these, that, those*) indicate person(s) or thing(s) being referred to. Example: "*These* are my colleagues."

Relative Pronouns
The relative pronouns (*who, whom, which, that*) actually perform two functions. In addition to replacing nouns, they also link and establish the

relationship between a dependent clause and the main clause in a sentence. In the following example, the main clause consists of everything before the relative pronoun *who*; the remainder of the sentence is the dependent clause.

> The coach decided who would play that day.

There is much confusion about the relative pronouns *which* and *that*. The general rule is that *which* is used with nonrestrictive clauses (i.e., clauses that can be left out without changing the overall meaning of the sentence). The pronoun *that* is used with restrictive clauses, or clauses that cannot be left out without changing the meaning of the sentence. Here are two examples:

> The new cars, which are more fuel efficient, are popular with consumers. (nonrestrictive clause)

> The marching band that led the parade was the best I've ever heard. (restrictive clause)

NOTE: As shown in the examples, nonrestrictive clauses are normally set off by commas, whereas restrictive clauses normally are not.

Indefinite Pronouns

Indefinite pronouns identify groups or classes of persons or things rather than a particular person or thing. Here are the most commonly used indefinite pronouns: *all, any, another, anyone, anybody, anything, both, each, either, everybody, few, many, most, much, neither, nobody, none, several, some,* and *such*. Example:

> A few voted against him, but everyone felt he would make a good chairman.

Interrogative Pronouns

The interrogative pronouns (*who, whom, what, which*) ask questions. They may introduce an independent interrogative sentence (direct question) or an indirect question included in a declarative sentence. Here are two examples:

What is going on here? (direct question)

He wondered who was in charge. (indirect question)

Reflexive Pronouns

The reflexive pronouns always end in—*self* or—*selves* and indicate that the subject of the sentence acts upon himself, herself, or itself. The reflexive pronouns are: *myself, yourself, himself, herself, itself, oneself, ourselves, yourselves,* and *themselves*. NOTE: *Myself* should not be substituted for *I* or *me* as a personal pronoun.

> Harold and myself made the sale. (incorrect)

> Harold and I made the sale. (correct)

Intensive Pronouns

The intensive pronouns are exactly the same as the reflexive pronouns, but their function is to provide emphasis to their antecedents (the pronoun that precedes and to which they refer). Example:

> We ourselves would like to know the answer.

Reciprocal Pronouns

The reciprocal pronouns (*each other* and *one another*) state the relationship between one item or person and another. Generally, *each other* is used when referring to two persons or things, while *one another* refers to more than two.

Common Usage Problems with Pronouns

Pronouns with Gerunds

Recall that a gerund is the "-ing" form of a verb that is being used as a noun. Only use the possessive form of pronouns with gerunds. Example:

> The manager was appreciative of him setting a new sales record. (incorrect)

The manager was appreciative of his setting a new sales record. (correct)

Case

In addition to the possessive "case," there are two other pronoun cases. Pronouns that are used as objects of verbs or prepositions are in the objective case (*me, us, him, her, it, you, them, whom*). Pronouns that are used as the subject of a sentence or clause are in the subjective case (*I, we, he, she, it, you, they, who*). Examples:

They loved their dog. (subject of the sentence, so subjective case)

The dog loved them. (object of the verb, so objective case)

Sometimes compound pronouns cause problems with regard to case (e.g., whether you should use *me* or *I*). To determine which case to use, mentally use the pronouns singly in the sentence. Example:

Jack phoned her and (me? I?).

Jack phoned . . . me. (correct)

Case can also be a problem when pronouns modify nouns. Mentally remove the noun to determine the case of the pronoun. Example:

(We? Us?) engineers use a standard set of symbols.

We use a standard set of symbols. (correct)

After the words *as* or *than*, pronoun case can be a question. To resolve the issue, add the words that have been left out of the sentence. Example:

Larry was a better athlete than (him? he?).

Larry was a better athlete than he (was). (correct)

Gender

Pronouns must agree in gender with their antecedents. Problems sometimes arise because masculine pronouns have traditionally been used to refer to both sexes. Example:

> Every member may vote as he chooses. (grammatically correct, but sexist)
>
> Every member may vote as he or she chooses. (better)
>
> All members may vote as they choose. (also acceptable)

Number

Number (i.e., singular versus plural) is sometimes a question with indefinite pronouns such as *each, either, neither, anybody, anyone, everybody, everyone, nobody, no one, somebody,* and *someone.* These pronouns are almost always singular. Thus, they require singular verb forms, and any pronouns that refer to them should also be singular. Example:

> Neither of the companies is listed on the New York Stock Exchange.

Memory Solidifier

Instructions: Under each of the headings below, paraphrase (i.e., write in your own words) the main points under that heading in the material you have just studied. If necessary, return to the previous material to refresh your memory.

Give the definition of a pronoun:

-

List the three groups of personal pronouns:

-
-
-

List the demonstrative pronouns:

-
-
-
-

List the relative pronouns:

-
-
-
-

List the interrogative pronouns:

-
-
-
-

Indefinite pronouns do what?

-

Reflexive pronouns always have one of two endings. They are:

-
-

What is the function of intensive pronouns?

-

List the two reciprocal pronouns:

-
-

List any points on pronoun usage that you feel you may have trouble remembering:

Parts of Speech (cont'd)

Verbs

Definition: A verb is a word or group of words that does one of the following:

- Describes an action: "The tight end *caught* the ball."
- Indicates how a person or thing is affected by an action: "They *were uplifted* by the sermon."
- Expresses a state of being: "My son's room *is* messy."

Types of Verbs

There are two types of verbs: transitive and intransitive. The latter include linking verbs.

Transitive Verbs

A transitive verb indicates that the subject of the sentence acts upon someone or something. Another way to say this is that a transitive verb always requires a direct object to complete its meaning. Example:

> Handel composed "The Messiah" in less than a month. ("The Messiah" is the direct object of the verb composed.)

Intransitive Verbs

An intransitive verb can make a complete statement about its subject by itself, and thus it does not require a direct object. (Intransitive verbs may, however, have modifiers such as adverbs.) Examples:

The car started.

The car started easily. ("easily" modifies the verb)

Some verbs are called linking verbs because they link a complement to the subject of the sentence. The following intransitive verbs are almost always used as linking verbs: *be, become, seem, appear*. Others, such as *look, sound, taste, smell,* and *feel*, can be either linking verbs or simple intransitive verbs.

If the complement of an intransitive verb is an adjective, it modifies the subject. Example:

The witness appeared agitated. ("agitated" modifies "witness")

Many complements are nouns or pronouns, which refer to, or restate, the same person or thing as the subject of the sentence. Example:

A hammer is a useful tool. ("tool" restates "hammer")

Forms of Verbs

There are two broad forms of verbs: finite and nonfinite.

Finite Verbs
A finite (or defined) verb can make a complete statement about its subject. Example:

The explosion broke our windows.

Finite verbs can change form to express person, tense, and number (see the topic "Properties of Verbs" below).

Nonfinite Verbs
Nonfinite verbs are also called verbals. They consist of verb forms, but they function as nouns, adjectives, or adverbs.

An infinitive (the "to" form of a verb) can be used as any of the above.

She loves to sing. (noun: direct object of loves)

> Now is the time to act. (adjective: modifies time)
>
> He dove to catch the ground ball. (adverb: modifies dove)

Recall that when the —*ing* form of a verb is used as a noun, it is a gerund. Example:

> Typing is her best skill. ("typing" is the subject of the sentence)

When verb forms other than the infinitive are used as adjectives, they are called participles. The *-ing* form of a verb is the present participle, and the *-ed* form is the past participle. Examples:

> The pianist was electrifying. (modifies pianist)
>
> Her prepared statement did not sway the board. (modifies statement).

Characteristics of Verbs

Most of the problems writers have with verbs are related to the many verb form changes required by the five characteristics of finite verbs: person, number, tense, voice, and mood.

Person

Just as personal pronouns have different forms for first, second, and third person singular and plural, so verbs change to agree in terms of person with their subjects. Example:

> George climbs (third person singular) mountains twice a year, but they climb (third person plural) every weekend.

This book does not cover all of the verb form changes related to person. When in doubt, consult a good dictionary or style guide.

Number

Verbs must also agree with their subjects in terms of number (i.e., singular or plural), and this can cause numerous problems. Some of the most common are discussed below.

- Some nouns have unusual or confusing singular or plural forms. Normally, the word *equipment* is considered singular, thus requiring a singular verb form. But what about *data*? The jury is still deliberating on this word, but because the singular form *datum* has almost entirely gone out of use the trend seems to be to use *data* as either a singular or plural noun, depending upon the context and the precise meaning intended. On this and other words, if you are uncertain, consult a company style guide or a general or specialized dictionary.
- Sometimes compound subjects joined by *and* can be confusing. Whether these subjects are singular or plural, they always require a plural verb.

  Bill and Marie keep their yards spotless. (they keep—third person plural)

- Singular subjects joined by *or* or *nor* require a singular verb.
  Neither rain nor snow delays the letter carrier. (rain delays—third person singular)
- If a singular and a plural subject are joined by *or* or *nor*, the verb should agree with the nearest to it.

  Neither government spending nor low interest rates have helped the Japanese economy to recover. (rates have—plural)

- When *and/or* is used to join two subjects (not a universally accepted practice), the question of number is ambiguous. The context should be used to decide if a singular or plural verb is needed.
- Collective nouns used as subjects may take either a singular or plural verb, depending upon whether the collective subject is being considered as one unit or as a group of individual members. Examples:

  The faculty is widely respected. (single unit, singular verb)

  The faculty disagree on the issue of tenure. (group of individuals, plural verb)

- When words such as *all, half, most, part,* or *some* are used, they are normally followed by an actual or implied prepositional phrase beginning with "of" and containing a noun. The verb form should agree in number with the noun in this phrase. Examples:

    Most (of the) problems arise because of inattention. (plural verb to agree with problems)

    Most of the land in Illinois is arable. (singular, to agree with land)

- When a number is used with a unit of measure to indicate a single measurement, use a singular verb. If (less common) the units of measure are considered individually, use a plural verb. Examples:

    In this circuit, 15 milliamps is the maximum allowable current. (one current level, singular verb)

    Up to two gallons are acceptable. (individual gallons, plural verb)

- Compound clauses will be discussed more fully later. For now, consider them to be a "list" within a sentence, each item of which is a clause with a subject and verb. In using compound clauses, writers sometimes leave out the "auxiliary" verbs (the verbs in the downstream list items). This is not necessarily a problem unless the subjects in the items vary with respect to number. Here is an example:

    The car was designed, the design approved, and the departments put to work on it. (change in number from car and design to departments)

The safest thing to do is to put in the auxiliary verbs in every list item:

   The car was designed, the design was approved, and the departments were put to work on it.

Tense

In addition to expressing an action or state of being, finite verbs also indicate the time that an action takes place or a state of being exists. Verbs do this through forms called tenses. In English, there are six tenses, as follows:

- Present tense (expresses an action or state of being in the present). Example: "He *paints*."
- Past tense (action or state of being that began and ended before the present). Example: "He *painted*."
- Future tense (action or state of being to begin in the future). Example: "He *will paint*."
- Present perfect tense (action or state of being that began in the past and is continuing in the present). Example: "He *has painted* for many years."
- Past perfect tense (action or state of being that began and ended in the past—often, simple past tense can be substituted for past perfect). Example: "He *had painted* earlier in his life."
- Future perfect tense (action or state of being that will begin after the present and end at some point in the future). Example: "He *will have painted* many canvasses by the time he lays down his brushes."

A complete study of tenses and tense sequences (the time relationships among the various verbs and verbals in sentences) is beyond the scope of this book. For full information, see a good grammar text or a style manual of the English language.

The things to remember are:

- Verb tenses should be appropriate for the time relationships being expressed.
- Tenses should not be shifted within sentences, paragraphs, or sections of a document unnecessarily, as this is very confusing to readers.

Voice

The concept of "voice" refers to the two forms of verbs that indicate whether the subject of the verb acts (active voice) or is acted upon (passive voice). The passive voice verb construction consists of a form of the verb *to be* followed by the past participle of another verb. Examples:

The workers fixed the leaking pipe under the street. (active voice)

The leaking pipe was fixed more than a week ago. (passive voice)

Use the active voice as much as possible in technical and business writing, as it is more forceful, more direct, and easier to understand. There are times, however, when the passive voice may be appropriate. Here are a few:

- When the actor is not mentioned or is unknown or unimportant (see example above, in which "more than a week ago" is more important than who fixed the pipe).
- When active voice is too abrupt, or a weaker imperative (command) is desired. Example:

  Follow her instructions immediately. (active voice)

  Her instructions should be followed immediately. (passive voice)

- When the receiver of the action requires emphasis. Example:

  Only the best students are admitted to that school. ("students" is emphasized, rather than "school")

Mood

The concept of "mood" means that verb forms can be used to express how the writer regards the statement being made. This concept is not as important in English as in some other languages. However, three moods are still distinguished: indicative, imperative, and subjunctive.

- The indicative mood is the form in which we make statements and ask questions. (Examples: "I am." "Is that true?")
- The imperative mood is the form used in commands and instructions. As such, it plays a major role in technical writing, as large portions of procedures contain lists of detailed instructions. (Example: "Install the lockwasher and the nut.")
- The subjunctive mood verb forms used in English are almost all made with forms of the verb *to be* combined with another verb.

The subjunctive has almost passed out of English usage, except for two situations, as follows:

- Repeating or reiterating a command, suggestion, or requirement. (Example: "He requires that we be seated by eight.")
- Indicating a condition that is a wish or that is contrary to reality, doubtful, improbable, or unlikely to occur. (Example: "If our aggressive impulses were controlled, the world would be a better place.")

Memory Solidifier

Instructions: Under each of the headings below, paraphrase (i.e., write in your own words) the main points under that heading in the material you have just studied. If necessary, return to the previous material to refresh your memory.

A verb is a word that does one of three things. What are they?

-
-
-

List and define two "types" of verbs:

-
-

Define linking verbs:

-

List and define two "forms" of verbs:

-
-

Define an infinitive, a present participle, and a past participle:

-
-
-

Define the "person" property of finite verbs:

-

List the six "tenses" of finite verbs:

-
-
-
-
-
-

Define active voice and passive voice:

-
-

Which voice should be used whenever possible?

-

Define the three "moods" of finite verbs:

-
-
-

Parts of Speech (cont'd)

Adjectives

Definition: Adjectives give additional information about nouns (i.e., subjects, direct objects, indirect objects), in one of two ways. They either point out a quality of the noun (descriptive adjective) or define the noun's boundaries (limiting adjective). Examples:

The tall tree (descriptive)

The night was dark (descriptive)

Four dollars (limiting)

Her dress (limiting)

Adjective Placement

Note that adjectives can appear either before the nouns they modify (called the attributive position), or after the nouns (the predicative position). In the latter case, the adjective normally follows a linking verb and is thus called a predicate adjective. Examples:

He likes driving his new car. (attributive, modifies car)

Her beauty is classic. (predicate adjective, modifies beauty).

Limiting Adjectives

The limiting adjectives include numerous categories of words, including many of the pronoun types that can also serve as adjectives.

- Articles: *a, an, the*
- Demonstrative adjectives: *this, that, these, those*
- Indefinite adjectives: *any, all, none, some, etc.*
- Interrogative and relative adjectives: *whose, which, that*
- Number adjectives: *(two, first)*
- Possessive adjectives: *his, her, my, our, their, your* (and plurals)

Degrees of Comparison

The three degrees of comparison for adjectives are the positive form (e.g., *calm*), the comparative form (e.g., *calmer*), and the superlative form (e.g., *calmest*). The comparative and superlative forms of most adjectives are formed in the same way (i.e., with the *-er* and *-est* suffixes). However, adjectives of two or more syllables are often preceded by *more* and *most* to form the comparative and superlative. Example:

> He thought Rome was (more/the most) beautiful.

Using Adjectives

Here are some common problems writers have with adjectives:

- *Adjectives with acronyms.* Should an acronym or abbreviation be preceded by *a* or *an*? Normally, the indefinite article *a* precedes a word beginning with a sounded consonant, while the indefinite article *an* precedes a word beginning with a vowel sound. With acronyms, however, the way the acronym may be expected to be read is the guide. Thus:

  He earned an MBA degree.

  He helped prepare a NASA proposal.

- *Omission of articles.* There has been a trend in much modern writing, especially technical writing, to omit articles for brevity. There is much discussion of this issue, but here are the most widely accepted guidelines:

  - Omission of articles is acceptable in technical lists set off from body text and in titles, headings, and figure captions.
  - For all other technical and business writing purposes, include the articles. The amount of brevity lost is insignificant, and the possibility of misunderstanding on the part of the reader is largely eliminated.

- *Multiple adjectives.* When more than one adjective modifies the same noun, separate the adjectives by commas, but do not put a comma between the last adjective and the noun. Example:

  The company built a robust, efficient, cost-effective mechanism.

- *Unit modifiers.* A unit modifier is a combination of two or more words that modify another word. Because of the detailed nature of technical and business writing, writers often find it necessary to draft nouns to serve as adjectives, and sometimes they line them up in cumbersome, hard-to-read "strings." When this happens, try rewriting the sentence using a prepositional phrase. Examples:

  Perform the electronic flow monitor voltage test.

  Perform the voltage test of the electronic flow monitor.

- *Hyphenation.* The question of whether or not to use hyphens with unit modifiers and other adjective constructions is a difficult one, requiring ongoing decision making and head scratching. Keep in mind the goal of all technical and business writing: absolute clarity. With this in mind, here are some guidelines for hyphenation:

  - **Do** use hyphens between the adjectives in the following circumstances:

- When the unit modifier contains numbers (example: 60-watt bulb).
- When a connecting word such as "and" is implied between the adjectives (example: They called it the Skinner-Peterson theorem).
- When the unit modifier contains a present or past participle (examples: light-emitting diode, well-tested design).
- Any situation in which the hyphen increases visual readability or ease of comprehension for the reader or contributes to clarity of communication.

- **Do not** use hyphens in the following circumstances:

  - When the first word is an adverb ending in -ly (example: newly designed part).
  - When a unit modifier is a proper noun (example: Federal Reserve Board Chairman).
  - When a unit modifier is a phrase adopted from another language (example: a priori condition).
  - With regard to predicate adjectives (adjectives following linking verbs), do not hyphenate a unit modifier unless it is hyphenated in the dictionary. Examples:

  The assembly was field tested.

  As an author, she is well-known.

- When the first word of the unit modifier is a comparative or superlative form (example: a more orderly process).
- When the unit modifier is a scientific name of a chemical, plant, or animal that is not normally hyphenated (example: a pure carbon dioxide atmosphere).

Memory Solidifier

Instructions: Under each of the headings below, paraphrase (i.e., write in your own words) the main points under that heading in the material you have just studied. If necessary, return to the previous material to refresh your memory.

Define two types of adjectives:

-
-

Adjectives may be placed where relative to the nouns they modify?

-

List the six types of limiting adjectives:

-
-
-
-
-
-

List the three degrees of comparison:

-
-
-

What is the rule for using *a/an* with acronyms or abbreviations?

-

What is the current thinking about omitting articles in technical and business writing?

-

What is the rule for punctuating multiple adjectives that modify the same noun?

-

What is a unit modifier?

-

List any rules for hyphenation that you think you may have trouble remembering:

Parts of Speech (cont'd)

Adverbs

Definition: Most adverbs are adjectives or participles with *-ly* endings (examples: *lastly, despairingly*). They modify verbs, adjectives, and other adverbs (but not nouns or pronouns). They can also modify entire sentences, connect and modify clauses, and introduce questions. Here are some examples:

He performed recently. (modifies the verb performed)

She is very intelligent. (modifies adjective intelligent)

They won almost effortlessly. (adverb "almost" modifies adverb effortlessly)

Unfortunately, we could do nothing. (modifies sentence)

It was done; however, they could not accept it. (connects two clauses and modifies their meaning)

When was the last time you saw him? (introduces a question)

Another way to define adverbs is to say that they answer one of the following questions: How? How much? When? Where?

Adverbials

There are three other types of constructions that are used in adverbial functions: nouns, adverbial phrases, and clauses used as adverbs. Examples:

> He works nights. (nights is a noun modifying the verb works)

> After lengthy experiments, they succeeded. (phrase modifying the verb succeeded)

> He scored well on the test, because he had studied. (clause, with its own subject and verb, modifying entire first clause)

Types of Adverbs

There are four types of adverbs, as follows:

- *Common adverbs.* This category includes almost all adverbs. Examples: *inherently, almost, blindly.*
- *Conjunctive adverbs.* These adverbs serve joining functions in sentences. They include such adverbs as *accordingly, consequently, however, nonetheless, therefore.*
- *Interrogative adverbs.* These adverbs ask questions. They include *how, when, where, why, who.*
- *Numeral adverbs.* The most common of these are *once* and *twice*, but words such as *often* and *seldom* are sometimes put in this category.

Usage Problems with Adverbs

Adverb Placement

Adverbs can be positioned almost anywhere relative to other sentence elements, so there are no completely general guidelines in this area. However, adverb placement can cause problems of meaning, so the best advice is to be on the lookout for ambiguous or erroneous meanings caused by adverb positioning. Here is an example of what is called a squinting adverb, because it is unclear whether it modifies what comes before or after it.

> While stirring the compound slowly pour in the granules.
> (ambiguous)
>
> While stirring the compound, slowly pour in the granules.
> (corrected with punctuation)
>
> While slowly stirring the compound pour in the granules.
> (corrected with different adverb placement)

Adverbs such as *also, almost, nearly, only,* etc. should be placed as close as possible to the words they modify in order to avoid ambiguity. Example:

> The assumption is only proved if X = 8.633.

Do you want to say this, or can the assumption be proved in other ways? Perhaps what you want to say is this:

> The assumption is proved only if X = 8.633. (this limits the statement to functions of X but does not preclude other proofs of the assumption)

Split Infinitives

Recall that an infinitive is the "to" form of a verb. When something is placed between the "to" and the verb it is called a split infinitive. Traditionally, split infinitives have been looked down upon in formal writing, but it is universally agreed that it is acceptable to split an infinitive (usually with an adverb) to prevent ambiguity of meaning or awkward sentence structure. Example:

> He decided quickly to change the valve stem mechanism.
> (ambiguous)
>
> He decided to change quickly the valve stem mechanism.
> (awkward)

He decided to change the valve stem mechanism quickly. (possibly OK, but potentially ambiguous since the adverb is far from the two verbs it might modify)

He decided to quickly change the valve stem mechanism. (uses a split infinitive, but avoids ambiguity and is less awkward)

Memory Solidifier

Instructions: Under each of the headings below, paraphrase (i.e., write in your own words) the main points under that heading in the material you have just studied. If necessary, return to the previous material to refresh your memory.

List the three types of words that adverbs modify:

-
-
-

What are the four questions that adverbs ask?

-
-
-
-

What are adverbials?

-

List the four types of adverbs:

-
-
-
-

Where can adverbs be placed relative to the words they modify?

-

What is a "squinting" adverb?

-

Adverbs such as *also, almost, nearly,* and *only* should be placed where relative to the words they modify?

-

What is a split infinitive, and when is it acceptable to use one?

-
-

Parts of Speech (cont'd)

Prepositions

Definition: A preposition is a word that indicates a relationship between a noun or pronoun (called the object of the preposition) and another word or element in a sentence. This relationship can be one of physical location (e.g., *in, between, beyond, under*), abstract orientation (*for, against, contrary to*), time (*after, during, until*), or direction (*into, through, toward*).

A preposition, with it its object (a noun or pronoun) and any modifiers of the object, form a *prepositional phrase.*

Linguistic Functions

Functionally, prepositional phrases serve as modifiers of other words or sentence elements. They can serve as:

- Adjectives modifying nouns or pronouns.
- Adverbs modifying verbs, other adverbs, or adjectives.

When serving as an adjective, a prepositional phrase always follows the noun or pronoun it modifies. Here is an example:

> The topic of our conversation was politics. (adjective, modifying the noun "topic.")

When serving as an adverb, a prepositional phrase can appear in various places relative to the word or sentence element it modifies. Examples:

She put the flowers on the table. (adverb, modifying the verb "put.")

Of their accomplishments, they were understandably proud. (adverb, modifying adjective "proud.")

Many words that are used as prepositions also function as other parts of speech, especially adverbs. For example:

The boulder rolled down the hill. (preposition)

The lead runner fell down, after developing a cramp in his leg. (adverb)

Common Usage Problems

Do not use redundant forms such as "off of" or "at about." At the same time, do not omit prepositions where they are needed. Example:

The pastor took care and was loved by his congregation. (incomplete, confusing)

The pastor took care of and was loved by his congregation. (correct)

If a preposition falls naturally at the end of a sentence, it is grammatical to leave it there. However, this usually results in a weak ending for the sentence. The phrases "in which" and "into which" can still be used, though often this results in awkward constructions. The other alternative is to rewrite the sentence to avoid either of these two results. Examples:

He could not remember the drawer he had put his car keys in. (OK, but weak)

He could not remember the drawer into which he had put his car keys. (OK, but awkward)

He knew his car keys were in a drawer, but he could not remember which one. (longer, but stronger)

Certain verbs, adverbs, and adjectives are used with prepositions to form what are called *idiomatic expressions*. There are no formal grammatical rules for determining which prepositions are correct for particular constructions of this type (that is why they are called idiomatic), but usage has dictated the appropriate form in almost all cases. Below are just a few idiomatic expressions involving prepositions. For others, consult your dictionary.

analogous to   different from   implicit in   inconsistent with

In titles, capitalize prepositions only if they are the first word (do not use a preposition as the last word) or if they are *four or more* letters long (e.g., "Waiting *for* Lefty," "Come Fly *With* Me").

Memory Solidifier

Instructions: Under each of the headings below, paraphrase (i.e., write in your own words) the main points under that heading in the material you have just studied. If necessary, return to the previous material to refresh your memory.

Give the definition of a preposition:

-

Prepositions express four types of relationships between the object of the preposition and other sentence elements. These relationships are:

-
-
-
-

The object of a preposition must be either:

-
-

What is a prepositional phrase?

-

What are the two linguistic functions served by prepositional phrases?

-
-

When serving as an adjective, the prepositional phrase should be placed where relative to the word it modifies?

-

When serving as an adverb, the prepositional phrase can be placed where relative to the word it modifies?

-

What is an idiomatic expression?

-

In titles, prepositions should be capitalized in what two cases?

-
-

Parts of Speech (cont'd)

Conjunctions

Definition: Conjunctions are words that connect sentence elements and indicate the relationship between the words, phrases, or clauses being connected. For such a small group of words, the classification of conjunctions will probably seem like "terminology soup" at first. However, it is very logical, and the different types of conjunctions need to be understood if you are going to use them correctly. Let's take it step by step.

Coordinating Conjunctions

Some conjunctions join sentence elements of *equal grammatical rank* (i.e., they have the same function). As mentioned above, these elements can be words, phrases, or clauses. Conjunctions of this type are broadly classified as coordinating conjunctions because they coordinate between equal sentence elements. Within the category of coordinating conjunctions there are three subcategories, as follows.

Standard Coordinating Conjunctions
Standard coordinating conjunctions (sometimes called coordinate conjunctions) can join words, phrases, and clauses of equal rank. They include *and, but, so, for, nor, or,* and *yet.* Examples:

> The Baroque and Classical Periods preceded the Romantic Period in musical history. (connecting two adjectives)

> Should we vacation in the city or at the lake? (connecting two prepositional phrases)

> He applied the voltage, but nothing happened. (connecting two nonrestrictive or "independent" clauses)

In using standard coordinating conjunctions, remember that they must join sentence elements of equal rank. For example, do not use them to join a noun and a phrase or a noun and a clause.

> Clean the front, sides, and in the corners of the unit. (wrong)

> Clean the front, sides, and corners of the unit. (correct)

> He loved his work and that the company appreciated his efforts. (wrong)

> He loved his work and the fact that the company appreciated his efforts. (correct)

Correlative Conjunctions

Correlative conjunctions are coordinating conjunctions that come in pairs (e.g., *either . . . or, neither . . . nor, not only . . . but also, both . . . and,* and *whether . . . or*). They "correlate" two things in a sentence. Here are two rules:

- Each part of a correlative conjunction must be followed by the same part of speech.
- It is best to maintain strict "parallelism" in sentence structure when using correlative conjunctions.

Here are examples to show what we mean by these rules:

> The Engineering Department performed both design functions and documented its designs. (differing parts of speech—noun with modifier vs. verb)

> The Engineering Department performed both design and documentation functions. (same part of speech—adjectives)

> The spacecraft onboard computer either activated the science instruments or could place the system in a safe-hold condition.

(nonparallel construction in sentence elements—in this case verb forms)

The spacecraft onboard computer could either activate the science instruments or place the system in a safe-hold condition.

Conjunctive Adverbs

Conjunctive adverbs are adverbs that function as conjunctions because they join two nonrestrictive (independent) clauses. The most common conjunctive adverbs are: *accordingly, also, besides, consequently, further, furthermore, however, likewise, moreover, namely, nevertheless, then,* and *therefore.* Here is the rule for punctuating conjunctive adverbs:

- Conjunctive adverbs most often are preceded by either a semicolon or a period and must be followed by a comma. Examples:

    There is always time to do things over, but there is never enough time to do them right the first time. (using a coordinating conjunction)

    There is always time to do things over; however, there is never enough time to do them right the first time. (conjunctive adverb with semicolon)

    There is always time to do things over. However, there is never enough time to do them right the first time. (conjunctive adverb with period)

    There is always time to do things over, however there . . . (wrong!)

Subordinating Conjunctions

The second broad category of conjunctions is subordinating conjunctions. As the name implies, these conjunctions are words that connect sentence elements of unequal rank, one subordinate to the other. In other words, subordinating conjunctions connect restrictive or dependent clauses to nonrestrictive or independent clauses. This category includes the relative pronouns *who, whom, whose, which,* and *that,* plus the following (among others): *after, although, as, before, if, since, than, that, though, unless, when, where,* and *whereas*. Here are some rules for using subordinating conjunctions:

- Use *if* to indicate condition. Use *whether* to indicate alternatives.

    Wait to see if the LED comes on.

    Monitor the gauge to determine whether the level is above or below the specified limit.

- Remember that the relative pronoun/conjunction *that*, with no comma preceding it, should be used to connect a restrictive clause to a nonrestrictive clause. Use *which*, preceded by a comma, to connect nonrestrictive clauses. Examples:

    He caught the pass that the quarterback threw just before being sacked. (the second part of the sentence restricts, or completes, the meaning of the first part)

    He made an interception, which he had never done before. (the two clauses are nonrestrictive and can stand alone)

- After verbs such as *believe, say,* and *think*, the subordinating conjunction *that* may sometimes be omitted, but to guarantee absolute clarity this practice is not recommended. Examples:

    I believe the statement was made in jest. (possibly confusing)

    I believe that the statement was made in jest. (unambiguous)

- Do not repeat the word *that* when a phrase or clause comes between it and the sentence element it is connecting to the rest of the sentence. Here is an example:

    She suspected that due to the holiday that the Post Office would not be open. (incorrect)

She suspected that due to the holiday the Post Office would not be open. (correct)

NOTE: One way to clarify your thinking in such situations is to mentally or physically enclose the intervening material in commas:

She suspected that, due to the holiday, the Post Office would not be open. (also correct)

Memory Solidifier

Instructions: Under each of the headings below, paraphrase (i.e., write in your own words) the main points under that heading in the material you have just studied. If necessary, return to the previous material to refresh your memory.

Give the definition of conjunctions:

-

Define coordinating conjunctions and list some examples:

-

Define correlative conjunctions and list some examples:

-
-

List two rules for using correlative conjunctions:

-
-

Define conjunctive adverbs and list some examples:

-

What is the rule for punctuating conjunctive adverbs?

-

Define a subordinating conjunction and list some examples:

-
-

List any of the usage rules for subordinating conjunctions that you think you may have trouble remembering:

Sentence Structure

Introduction

Sentence structure is a complex topic, but it is at the very center of your ability to be a good writer. The parts of speech do not communicate until you craft them into good English sentences. If you desire to be a good writer in English, you must make it your life's work to understand English sentence structure.

In this book there is not time to teach you everything about sentence structure, but many books, such as general style guides, have excellent discussions of all topics related to sentences and sentence structure. As time goes on, you may wish to study these topics further.

Here are four very general principles of good sentence structure:

- Put your main idea in the main clause of the sentence.
- Try to use "simple sentence structure" (subject, verb, direct object) as much as possible. (Note: This does not preclude using more complex sentences.)
- Use parallel construction.
- Avoid sentence fragments.

We will discuss these principles more fully, but first let's look at the basics of sentence construction.

Types of Sentences

There are four types of sentences: simple, compound, complex, and compound-complex. Let's examine each of them.

Simple Sentences

A simple sentence states one main idea, and it has only a main clause with no subordinate clauses. Here is a simple sentence:

> He brought her flowers.

In this sentence, "he" is the subject, "brought" is the verb, "her" is the direct object, and "flowers" is the indirect object. It is a clear, direct, and forceful sentence.

Compound Sentences

Compound sentences contain two or more complete ideas (i.e., two or more main clauses) with no subordinate clauses. Most often, the main clauses are joined by coordinating conjunctions, with a comma preceding the conjunction. They can also be joined by a semicolon, although in technical and business writing we tend to avoid the use of semicolons because they are imprecise and are often misused. Here are examples of compound sentences:

> The failure feedback circuit alerted the operator with an error message, and it prevented the mechanism from moving.

> The signal was sent, but the spacecraft antenna failed to deploy.

Notice that the clauses on each side of the comma have subjects and verbs. In other words, they could stand alone as sentences. For that reason, they are separated by punctuation. Note that it is possible to omit the comma in some very short compound sentences, particularly when the clauses are joined by the conjunction *and*, but this practice is not recommended, since there is potential for confusion.

Complex Sentences

Complex sentences are just that—complex. Think of complex sentences as having one main clause (which should contain the main idea of the sentence) and some extra material that modifies the main clause (i.e., one or more dependent clauses). Here is an example:

> Physicists will know whether the universe is going to collapse when they are able to measure the matter in the universe precisely.
>
> (Alternate construction) When they can measure the matter in the universe precisely, physicists will know whether the universe is going to collapse.

The clause "when they can . . ." is dependent. It adds information to the main clause "physicists will know . . . ," which could conceivably stand alone as a sentence (i.e., it has a subject and a verb). Notice that punctuation depends upon the placement of the subordinate clause. If it appears at the beginning of the sentence and is fairly long, set it off with a comma. If it appears after the main clause, it can often go without punctuational separation.

By the way, which of the sentences above is more direct and clear? The first, of course. It places the main idea at the *beginning* of the sentence, which gives the main idea prominence.

Compound-Complex Sentences

Alas, there is such a thing as a compound-complex sentence! This is a compound sentence (two or more main clauses) with at least one dependent clause that modifies one of the main clauses. Here is an example:

> The Federal Reserve Board can raise interest rates, which clamps down on the money supply, but consumer confidence is still a crucial factor in overall spending in the economy.

In this sentence, the clause before the first comma is an independent clause. The material between the commas is a dependent clause (it modifies the first main clause), and the material after the second comma is a second independent clause. Another way to state this is to say that you could leave out the material between the commas and have a solid compound sentence, with two main clauses.

Phrases

A "phrase" is a group of words that has *either* a subject or a finite verb, *but not both*. There are three kinds of phrases: prepositional phrases, noun phrases, and verb phrases.

Prepositional Phrases

You recall that prepositional phrases are introduced by prepositions. These phrases may serve various grammatical functions in sentences, but most often they are used to modify nouns (adjectival use) or to modify verbs (adverbial use). Here are some examples of prepositional phrases in sentences:

> At night the sky was filled with stars. (prepositional phrases "at night" and "with stars" both modify the verb "filled")

> The chairman of the committee cast the deciding vote. ("of the committee" is a prepositional phrase modifying the noun "chairman")

Noun Phrases

A noun phrase consists of a noun and its modifiers. Such phrases serve grammatically as nouns in sentences. Example:

> The beautiful flowers were placed in a vase on the table. ("The beautiful flowers" is a noun phrase that is the subject of the sentence)

Verb Phrases

There are actually three types of verb phrases: participial phrases, infinitive phrases, and gerund phrases. (Remember that gerunds are verb forms that function as nouns.)

Participial Phrases

A participial phrase consists of a participle (a verb form usually ending in *-ing*, *-ed*, or *-en*) and possibly also an object and modifiers. Participial phrases serve as adjectives, modifying nouns. Example:

> The mechanic doing the tune-up was very precise. ("doing the tune-up" is a participial phrase serving as an adjective that modifies the noun "mechanic")

Infinitive Phrases

An infinitive phrase is the "to" form of a verb, along with a modifier and possibly an object. Infinitive phrases can serve as nouns, adjectives, and adverbs. Here is an example of each usage:

> She wanted to win the promotion. ("to win the promotion" is an infinitive phrase serving as a noun and functioning as the direct object of the verb "wanted")

> His handout contained the ideas to be discussed. ("to be discussed" is an infinitive phrase serving as an adjective modifying the noun "ideas")

> The favored runner failed to complete the race. ("to complete the race" is an infinitive phrase serving as an adverb modifying the verb "failed")

Gerund Phrases

Gerund phrases consist of a gerund, a modifier, and possibly an object. They serve, like simple gerunds, as nouns. Here are two examples:

> His having been a manager before gave him the edge for the new job. ("His having been a manager before" is a gerund phrase that serves as the subject of the sentence)

> She loves travelling in Asia. ("travelling in Asia" is a gerund phrase and is the direct object of "loves")

Things to Remember About Clauses and Phrases

- Clauses have *both* a subject and a finite verb.
- Phrases have *either* a subject or a verb, *but not both*.
- Main clauses *can* stand alone as sentences.
- Subordinate clauses *cannot* stand alone.

Put Main Idea in Main Clause

Now that you are re-acquainted with sentence structure, we can talk about how to create better sentences.

Sentences are more forceful, clear, and direct if the main idea is contained in a main clause. Here are two ways of making a statement, one of which has stronger construction:

> Although George is the man we want, we can't get him. (weak construction)

> Although we can't get him, George is the man we want. (stronger)

Here is a sample procedural instruction:

> After inserting the screw, torque to 40 lbf•in. (weak)

> After insertion, torque the screw to 40 lbf•in. (stronger)

Use Simple Sentence Structure

The thought behind this is not that you should always use simple sentences. Indeed, varying your sentence length is a good way to keep your writing interesting and retain the reader's attention. However, **The Handbook of Technical Writing**, by Charles T. Brusaw, et al., says this:

> "Use uncomplicated sentences to state complex ideas. If readers must cope with a complicated sentence in addition to a complex idea, they are likely to become confused." (p. 615)

This concept is very important for business and technical writing, because *clarity and accuracy are the foremost goals*, with the beauty of the prose coming in a distant third.

Here is an example of an overly complicated sentence rewritten to make it simpler and more easy to understand:

> When we send parts out for machining several times during the production process, and especially when they must go to different vendors, some of whom cannot guarantee fast turnaround, we lose time and money, and we cannot satisfy our own customers' requirements. (grammatically correct, but too complicated)

> When we send parts out for machining several times during the production process, we lose time and money, and we cannot satisfy our own customers' requirements. This situation is especially critical when we must send the same parts to different vendors, some of whom cannot guarantee fast turnaround. (stronger: main ideas in first sentence, elaboration in second sentence)

It takes some practice to learn to communicate complex ideas in uncomplicated sentences, but every minute you think about your sentences will be time well spent.

Use Parallel Construction

Parallel structure means that sentence elements that are alike in function should be expressed with the same type of construction. Parallel structure can be achieved using words, phrases, or clauses. Here are some examples of good parallelism:

> The documentation package included schematic, assembly, and part locator drawings. (parallelism of words)

> The toolbox contained a crescent wrench, a pipe wrench, and a socket wrench. (parallelism of phrases)

An automobile needs oil, it needs fuel, and it needs coolant. (parallelism of clauses)

Correlative conjunctions (*either . . . or, neither . . . nor, not only . . . but also*) can be especially tricky. The rule of parallelism is the same, however. Follow both parts of a correlative conjunction with words, similar phrases, or similar clauses—don't use a mixture. Here is good parallelism using correlative conjunctions:

Neither the guards nor the center could keep him from making a basket.

We are not only going to Spain, but we are also going to live there.

In technical and business writing, lists are often important for conveying information in an organized way. To achieve parallel structure in lists, each item (word, phrase, or clause) *should begin with the same part of speech*. Example:

Jones Co. merger on hold for the time being.

Profit situation improving.

Give the Legal Department all of the Jones documents. (faulty parallelism)

Jones Co. merger . . .

Profit situation . . .

Jones documents should all go to the Legal Department.

In the final analysis, parallelism is not especially difficult to achieve. All it takes is a little extra attention to detail. However, it helps the reader immensely, because it provides "cues" as to how the elements of a sentence relate to the subject of the sentence (or how list items relate to one another). If a number of sentence elements have similar relationships with the subject, parallel structure will tell the reader that immediately.

Avoid Sentence Fragments

Let's end our discussion of sentence structure—and the basics of English—with an easy one. A sentence fragment is an incomplete sentence that is nonetheless punctuated as a sentence. Remember that a sentence must have at least a subject and a *finite* verb. Sometimes writers mistakenly think that a group of words with a *verbal* in it can stand alone as a sentence. Verbals, however, are verb forms serving nonverb functions. Here are some examples:

> To climb the mountain. (fragment: no subject, and no finite verb)
>
> They wanted to climb the mountain. (a good sentence; "to climb the mountain" is a verb phrase that serves as the direct object of "wanted")
>
> But not with him. (fragment: no subject or finite verb)
> She met with Joe, but not with him. (sentence)

Memory Solidifier

Instructions: Under each of the headings below, paraphrase (i.e., write in your own words) the main points under that heading in the material you have just studied. If necessary, return to the previous material to refresh your memory.

List four principles of good sentence structure:

-
-
-
-

What is the definition of a simple sentence?

-

Write a simple sentence:

-

What is the definition of a compound sentence?

-

Write a compound sentence:

-

What is the definition of a complex sentence?

-

Where should the main idea go in a complex sentence?

-

Write a complex sentence:

-

What is the definition of a compound-complex sentence?

-

Write a compound-complex sentence:

-

What is the definition of a phrase?

-

What is a prepositional phrase?

-

What are the two grammatical uses of prepositional phrases?

-
-

Write two sentences using prepositional phrases in the ways listed above.

-
-

What is a noun phrase, and what is its grammatical function in a sentence?

-
-

Write a sentence with a noun phrase:

-

List and define three types of verb phrases:

-
-
-

Write three sentences using verb phrases in the ways listed above:

-
-
-

List four things to be remembered about clauses and phrases:

-
-
-
-

Why should the main idea be placed in the main clause?

-

Why should you use simple sentence structure as much as possible?

-

What do we mean by parallel construction?

-

What is a sentence fragment?

-

Part 2

Technical Writing

Introduction

This section is not a comprehensive course in technical writing. As a working professional you are assumed to be familiar with the types of technical documentation that are used in business and industry. These include:

- Specifications
- Test procedures and acceptance test procedures
- Assembly and disassembly procedures
- Calibration procedures
- Maintenance manuals and troubleshooting procedures
- Operation manuals and procedures
- Software documentation.

The specific content of these documents varies across industries and companies, so there is little benefit in studying "generic" documents. You must become familiar with them at your place of employment.

There are, however, many valuable things that can be learned about doing good documentation, and this information generalizes very well across all types of documentation writing that you may be required to do. This section concentrates on such information, and particularly on areas that have historically been most problematic for working technical professionals.

The First Two Laws of Good Technical Writing

Do *even more* for your reader than you would require if you *were* your reader.

Do everything you can to make your documentation absolutely, totally, 100% "goof proof."

Short Forms

First Use

Spell out all terms on first use (except units of measure), with the acronym in parentheses following. Example:

> line replaceable unit (LRU).

It is also helpful to spell out terms at selected places throughout a document if usages of the terms are widely separated.

Initial Caps

It is not necessary to initial-cap every word in a spelled-out term that is represented by an all-cap acronym. Example:

> programmable read-only memory (PROM)

(The requirement to capitalize the acronym letters in the spelled-out term is an old rule that has simply become too difficult to maintain, given the explosion of acronyms in industry and elsewhere.)

Articles With Acronyms

To decide whether to use the article "a" or "an" before an acronym, speak the acronym aloud (not the spelled-out term). The phonetic sound of the acronym is the determining factor. Example:

> Figure 1 shows an SOV of the type used in this assembly.

Some acronyms, like "maser" and "LEM" can be pronounced like words. The guide is still how the acronym is generally read aloud. Thus: "The design called for a maser." "The astronauts descended to the Moon's surface in a LEM."

Units of Measure

When using abbreviated units of measure, there should be (with few exceptions) a space between the numeral and the unit designator. One exception is the % symbol, which can be run up against the numerals.

Degree Symbol

The degree symbol (°) is used to express quantities of various kinds, including temperature, longitude/latitude, and circumference. When it is used to express temperature, close up the symbol with the scale indicator, *not* with the numeral. Example:

80 °F.

When the symbol is used to express things other than temperature, close up the symbol with the numeral. Example:

39° N Latitude.

Plurals

Acronyms can be pluralized by adding a lower-case "s" (without apostrophe). Units of measure are NEVER pluralized.

English and Metric Units

Specifications, tolerances, etc., should be expressed in both English and metric units if the audience for the document is international.. The format is: English units (metric units).

Period With Units of Measure

Units of measure *do not* have periods following. The one exception is "in.", the abbreviation for "inches." This is because without the period it would be confused with the word "in". When it is used as part of a compound unit of measure, however, it does not need the period. Example: lbf•in (pound-force inch).

Standards

Standardize your usage of short forms by adhering to one of the existing industry standards. Two prominent ones are:

- ANSI/IEEE Standard 260, Standard Letter Symbols for Units of Measurement
- MIL-STD-12, Abbreviations for Use on Drawings, and in Specifications, Standards, and Technical Documents

Memory Solidifier

Instructions: Under each of the headings below, paraphrase (i.e., write in your own words) the main points under that heading in the material you have just studied. If necessary, return to the previous material to refresh your memory.

What are the first two laws of good technical writing?

-
-

Write down, in your own words, any points relating to the use of short forms that are new to you or that you feel you might have trouble remembering.

Grammar and Punctuation

Syntax and Usage

Comprised of/Comprises

The phrase "is comprised of" is grammatically incorrect. "Comprises" is correct. To avoid confusion, substitute "consists of," "is composed of," or "encompasses." Even better: "includes," or "has/have."

Data

The word "data" can be used as either a singular or plural noun. Both are now considered correct. The only concern is that your pronouns and verbs must agree with your selected usage and with each other.

> Wrong: "This data are what we relied on."

> Correct: "These data are . . ." or "This data is . . ."

Interface

The word "interface" is a noun. Do not use it as a verb. Also, beware of "verb-izing" other nouns.

Mathematical Operators

Put a space on either side of mathematical operator symbols (plus, minus, etc.), including the following: $<$ $=$ $>$ $\pm$ . . . The exceptions are stand-alone specifications of voltage (e.g., +5 V) and range statements such as -.25/+.04. In these cases, the operator signs are closed up with the numerals.

Noun Clusters

Whenever possible, do not use noun clusters of more than three nouns.

>Poor example: "The nose landing gear uplock attachment bolt is . . ."

>Better example: "The bolt that attaches the uplock to the nose landing gear is . . ."

When you cannot avoid noun clusters, do one of two things:

>Use hyphens to show relationships between nouns: "Main landing-gear water-spray deflector . . ."

>(or)

>Use full terminology at first use, then redefine using a simpler term, which is used thereafter: "Engage the ramp service door safety connector pin (the pin that holds the ramp service door, hereinafter referred to as the safety connector pin) before you do this procedure."

Numbers Beginning Sentences

Spell out a number that begins a sentence. If the number is large (say, three digits or above), try to rewrite the sentence to place the number in a different position.

Spelling Out Numbers

Use numerals at all times with units of measure. At other times, the general rule is that the numbers 1 through 9 are spelled out, while numbers 10 and above are expressed in numerals. The primary reason for this is that numerals are simply much easier to read from the page or on a computer screen (and therefore they are more "goof proof").

However, if a sentence has numeric citations dealing with the same subject matter that include numbers both below and above 10, use the same format for all numbers. Normally, this will be numerals, but if for some reason a one-digit number must be spelled out (e.g., it begins the sentence), then all numbers in the sentence should be spelled out.

> Rule of thumb: when in doubt, use numerals, because they are easier to read from the page.

Four-digit numbers are not demarcated with a comma in running text. Numbers five digits and above do have commas.

Do not separate numerals from their unit symbols at the ends of lines.

Prefixes

The general rule is that prefixes such as "non" and "multi" should be attached to the root word in all cases unless there is a doubling of the vowel or a tripling of consonants.

The officially sanctioned exceptions to this rule are words beginning with the prefixes "co, de, pre, pro" and "re" (example: reengineering).

Following these rules, however, often leads to the creation of words that are visually very hard to read. In such cases, it is better to break the rules and put a hyphen between the prefix and the root word.

> Example: "multiamplifier"

> Better: "multi-amplifier"

Which/That

Which introduces clauses that are nonrestrictive (i.e., if they are left out, the meaning of the main clause does not change). Example:

> The adhesive, which was applied in step 4 above (nonrestrictive), will set in two hours.

Note that nonrestrictive clauses are normally set off from the rest of the sentence by commas.

That introduces restrictive clauses (though it may also introduce nonrestrictive clauses). Example:

> The lawn mower that is broken is in the garage.

Note that the clause introduced by *that* restricts the subject ("mower") to a specific one. Also note that restrictive clauses are normally *not* set off by commas.

In/Into, On/Onto

Items are inserted *into* (not in) and installed *onto* (not on) other items.

Grammar and Punctuation (cont'd)

Capitalization

Capitalize Sparingly

Of all grammatical elements, capitalization is the most apt to get out of control.

Capitalize only when there is a valid reason for doing so, not simply to draw attention to a word, part name, etc.

Hardware Names

Standard industry practice is to use initial caps for part or component names *only* when they are accompanied by their part number. Hardware names that include the word "system," "subsystem," or "assembly" generally are capped. Examples:

> Navigation Subsystem
>
> Hydraulic System
>
> high-gain antenna

General Rules

For information on the general rules of capitalization, refer to a good general style guide.

Grammar and Punctuation (cont'd)

Punctuation in Technical Writing

The Hyphen

The hyphen (-) has many uses. The first is to join compound nouns, such as "spin-up" and "self-test," that are not run together. Note that often the decision about running together or hyphenating is based upon readability, which has two elements:

- Visual readability.
- Agreement with the general rules of English pronunciation. For example, the noun "set-up" without the hyphen would seem to be pronounced "seet-up."

The second use of the hyphen is to join compound or "unit" modifiers (two or more words that modify a noun). Examples:

heat-resistant material

out-of-tolerance condition

The exception is that when the first word in a compound modifier ends in "ly" the hyphen is not used. Example:

tightly specified part.

The third usage of the hyphen is to break words at the ends of lines in text, which we generally do not do in technical writing.

The fourth use is in extended numbers or identifiers such as "145-892-7058."

Finally, the hyphen is used, as mentioned above, to set off prefixes from their root words when there is a doubling of a vowel or a tripling of consonants, or simply for readability.

Note that the hyphen should *not* be used as a minus sign unless absolutely necessary.

"En" Dash

The "en" dash (–) has three uses. First, it is the separator used in statements of "range." Example:

50–80 °F

Second, it can be used judiciously in lists (such as parts lists) for entries where an additional bit of information is added to the main entry. Example:

Control Assembly Configuration 1A.

Note that there are spaces before and after the dash in this type of usage.

Third, in many typefaces the "en" dash is an acceptable substitute for the minus sign.

"Em" Dash

The "em" dash (—) is used exclusively to set off parenthetical material. Example:

Only one person—the company president—can authorize shipment.

Minus Sign

Regarding the minus sign, all modern word processing programs have symbol sets that include the minus sign. If you find that it is identical to the program's "en" dash, use either. If not, opt for the true minus sign from the symbol set.

Quotation Marks

Use quotation marks *sparingly*. Aside from direct quoting (which is rare in technical writing) they may be used to indicate with precision how a term will be used in a document. Example:

> The Brake Control Unit, hereinafter referred to as "the control unit..."

Periods and commas are placed *inside* the quotation marks, semicolons and colons *outside*. Question marks, when needed, fall inside the quote marks if they are part of the material enclosed, but outside if they are part of the sentence as a whole. The main exception to these rules is that when a very short term or item is enclosed in quotation marks at the end of a sentence the period may fall outside the quote marks.

Serial Comma

In almost every style guide you will find the rule that a comma is not necessary after the next-to-last item in a series *unless there is a chance of confusion of meaning*. Ah, there's the rub! In technical writing we are concerned above all else with clarity and precision. Thus, make a habit of *always* using the discretionary (or "serial") comma between the last two items of a series. Example:

> Press the pushbutton, check that the LED comes on, and monitor the voltmeter reading.

Semicolon

If one or more items in a series contains punctuation (generally commas), separate the items with semicolons. Example:

> Remove the nut; the washer, lockwasher, and wire connector; and the bracket assembly.

Note that sometimes the list can be re-ordered so that the item with the internal punctuation can be placed last, making the use of semicolons for separators unnecessary. Except for separating list items (when necessary), *do not* use semicolons in technical writing.

Memory Solidifier

Instructions: Under each of the headings below, paraphrase (i.e., write in your own words) the main points under that heading in the material you have just studied. If necessary, return to the previous material to refresh your memory.

Write down, in your own words, any points related to grammar and syntax that are new to you or that you feel you might have trouble remembering.

-
-
-
-
-
-
-
-
-
-

Nomenclature

Clarity and Comprehension

When writing about hardware, it is very important to make sure that you are understood. Failure to communicate can result in poor assembly/disassembly, a faulty calibration, or a misapplied test procedure.

Capitalization

To reiterate, when hardware items have official names (systems, subsystems, assemblies, brand-name tools and equipment, etc.), the names should be initial-capped. Other hardware items that are designated only by common names, such as "locknut," are not capitalized.

Controls and Indicators

The nomenclature for controls and indicators should be *exactly* as placarded on the equipment. When a multi-position switch is not identified by a functional name (e.g., Power switch) and there are only two or three positions, identify it by the labeled positions (e.g., INT/EXT/OFF switch). When there are four or more switch positions, give the switch a functional name and use initial caps (e.g., Rate Count switch).

Switches

Switches (except for momentary pushbutton, détente pushbutton, and return-to-center toggle switches) are "set to" or "set at" a given position. Examples:

Set the Voltage switch to 28 V.

Check to see that all switches are set at RUN.

Set the Power switch to the ON position.

Pushbutton switches are pressed, not depressed. Example:

Press the SYNC switch and release.

Return-to-center toggle switches are held and released. Example:

Hold the MET switch in the CALIB position for three seconds, then release.

Pots and Selector Switches

Potentiometers and dials are "adjusted." Selector switches are "rotated" or "turned" to a position.

Lights and LEDs

Use the terms "come on" and "go off" to describe indicator functions, lights, and LEDs. Example:

The Power indicator will come on.

Readouts and Meters

Readout indicators "display" information. Meters "indicate" a scaled reading.

Verify That

When a response must be verified, the word "that" should be used after the word "verify". Example:

Verify the amount of flexure is within specified limits. (poor)

Verify that the proper voltage has been achieved. (better)

Memory Solidifier

Instructions: Under each of the headings below, paraphrase (i.e., write in your own words) the main points under that heading in the material you have just studied. If necessary, return to the previous material to refresh your memory.

Write down, in your own words, any points related to the use of nomenclature that are new to you or that you feel you might have trouble remembering.

-
-
-
-
-
-
-
-
-

Technical Writing Style

Active Voice

Write sentences in the *active voice* whenever possible. Active voice is preferred because it is direct and unambiguous. In the active voice, the subject of the sentence completes or performs the action. Example:

The control unit sends the command.

In the passive voice, the subject of the sentence is acted upon, and the "actor" is not the subject. Example:

The command is sent by the control unit.

The only times when the passive voice is preferred are when the action or the object of the action are more important than the "actor." Example:

The leakage problem was finally solved by a design change.

And/Or

Try to avoid using "and/or" in technical writing. On the surface it seems clear, but in fact it is vulnerable to misinterpretation. Why take the risk?

Clarity and Conciseness

Technical writing style should be *clear, direct,* and *concise,* and it should employ *good sentence structure.*

Introductory Clause

If you start an instruction with a dependent clause (e.g., in an "if/then" step), set off the introductory clause with a comma. Example:

> If there are no error indications, set the power switch to OFF.

"It" as a Subject

Avoid using "it" as a sentence subject referring to hardware, instrumentation, etc. There is great risk of ambiguous reference.

One Action per Procedural Step

Try to limit each procedural step to *one action, and perhaps its result.*

Remember that the person performing the procedure must be able to read a step, hold it in mind, look away, and do something. Steps that are too complex are vulnerable to misunderstanding or misapplication. Also, if you put more than one action in a step, you take the risk that the user will return to the procedure after doing the first action, think that the step is complete, and go on to the next step.

Procedural Steps

For procedural steps, use the *second person imperative mode, present tense.* (Or, to state it differently, tell the reader what to do in straightforward, simple terms.)

Recall that "second person" is the form of verbs that is used when the subject is "you." In the imperative mode, the "you" is generally implied, rather than stated.

In the imperative mode, we express a command, instruction, suggestion, request, etc. Examples:

> Install the unit into the test fixture.

> Turn the rotary switch to +15 V.

Specific Verbs

Use a specific action verb to describe an action, rather than a nonspecific verb or another part of speech. Examples:

> The meter gives an indication . . . . (poor)

> The meter indicates . . . (better)

Starting Sentences

Do not start a sentence with a symbol, numeral, abbreviation, or acronym.

State Actions Specifically

In writing procedural steps, state clearly and specifically the actions to be taken. Do not use abstractions or vague instructions. Examples:

> Ground the unit. (too vague)

> Attach a braided ground strap from earth potential to the unit's chassis. (better)

Telegraphic Style

In "telegraphic style," articles and other short words are left out in order to shorten sentences. In the past, this style was quite popular in technical writing, but today the pendulum has swung in the other direction. Good technical writers realize that every word in a sentence conveys meaning to the reader. For example, using the phrase "install *the* lock washer" rather than just "install lock washer" tells the reader that there is *only one* lock washer involved. If the shortened version is interpreted as "install *a* lock washer" it could imply that there are more washers to be installed or that more than one may be installed. In cases where manufacturing tolerances are tight, installing the wrong number of washers could cause severe problems.

If you do choose to use telegraphic style, use it *very sparingly*, employing utmost caution to ensure that the meaning is not changed by the omission of words.

Memory Solidifier

Instructions: Under each of the headings below, paraphrase (i.e., write in your own words) the main points under that heading in the material you have just studied. If necessary, return to the previous material to refresh your memory.

Write down, in your own words, any points related to technical writing style that are new to you or that you feel you might have trouble remembering.

-
-
-
-
-
-
-
-
-

Format

Standard Formats

There are two widely used standard formats for documentation: (1) military decimal, and (2) alphanumeric. Variations of both formats are also used.

Each type of formatting has benefits and weaknesses. Choose a format that best meets your needs, and use it *consistently*.

Military Decimal Format

Here is an example of the military decimal format:

> 1. 1.   *SECTION TITLE*
>
> 1. 1.1   *This is a 1st-level heading.* (There can be text following.)
>
> 2. 1.1.1 *This is a 2nd-level heading.* This is text following the heading. Note that the text wraps back to the left margin.
>
> 1.1.2   *This is another 2nd-level heading.*
>
> 1.1.2.1 *This is a 3rd-level heading.* This is text following.

The benefits of the military decimal format are that every paragraph has a heading and every heading is numbered. This means that every paragraph can be referenced in other parts of the text. It also means that the reader can locate all, or almost all, paragraphs by consulting the table of contents.

The weaknesses of the military decimal format are that it is cumbersome, hard to read, and confusing with regard to subordination of headings.

Note that there are modified versions of the military decimal format. In one variation, every paragraph is numbered, but not every paragraph is required to have a heading. In another variation, only headings are numbered, and paragraphs are not required to have a heading.

All decimal formats use the "flush-left" style.

Alphanumeric Format

Here is an example of alphanumeric formatting:

I. I. THIS IS A SECTION TITLE

A. This is a 1st-level Heading

    1. This is a 2nd-level Heading

       This is text following a 2nd-level heading. Text may follow any heading level. The text is indented to align with the heading and is "blocked" on the left (also called a "hanging indent").

       a. This is a 3rd-level heading, or it might be a "list item"—that is, one of a series of paragraphs or procedural steps.
       b. This is a second "list item."

The alphanumeric format is distinguished from decimal formats in that it uses alphanumeric identifiers rather decimal numbers. More importantly, it uses visual cues (indentions) to indicate levels of subordination. This is a very appealing feature of this type of format.

As with the military decimal format, there are variations of the standard alphanumeric format.

The principal problem that arises with the alphanumeric format is that when there are numerous levels of subordination the paragraphs move farther and farther to the right. To avoid this problem, reorganize the material to include a few more higher-level headings in order to bring the text back to the left.

Typefaces and Fonts

For running text on the written page, *serif* typefaces such as Times New Roman are easier to read than *sans serif* typefaces such as Arial or Helvetica.

Sans serif typefaces are good for titles and headings, especially in boldface. They are also acceptable for viewgraphs, posters, etc., where the font is usually large and prominent, and for computer text.

(NOTE: Times New Roman and Helvetica are "typefaces." A "font" is a particular typeface in a specified point size and style—e.g., Garamond 20 pt. bold).

Unless there are special circumstances, do not use more than *two* typefaces in a document (e.g., one for titles and headings and one for text).

Line Length and Margins

Longer lines of text are harder to read than shorter lines. Do not make your text lines longer than 6.5 inches. A 6-inch line is even better. This means that your left and right margins should be at least 1 inch. Margins of 1.25 inches are better, since they shorten the text lines and give a better binding edge.

Top and bottom margins should be 1 inch minimum, although page numbers and running footers can be 0.5 inch from the bottom of the page.

Justification

Text is easier to read if the right edge is left "ragged" (i.e., not justified), rather than being fully justified. Granted, from a distance fully justified type looks very nice, but the problem is that the spaces between words vary, which makes the text harder for the reader's eye to follow.

Table Format

Standard table format consists of heavy rules (e.g., 1.5 pt.) for the outside border and the line under the column heads, with lighter rules (e.g. 0.5 pt.) separating the rows and columns internally. Always provide internal rules, as they make table material 100 percent more readable. All "NOTES" and footnotes go outside and below the table.

Tables that are between half-page size and full-page size should be placed either at the top or bottom of the page. Smaller tables may be placed between paragraphs. Table titles are generally initial capped and boldface (i.e., title format) and appear centered above the table. Column headings are boldface, initial-capped, centered over the column, and "stacked" from the bottom.

Cell contents may be arranged in several ways. If the cell contents are verbal text, they should be flush left. If the cell contents are numbers of various sizes, they should be generally centered and justified on the right down the column. If they are numbers with decimal points, they should be generally centered and lined up on the decimal point down the column. If a cell has more than one line of data or text in it, data in other cells in the same row should be aligned with the top row of the multi-row cell. Here is an example of a good table:

TABLE 1. TABLE TITLE HERE

Column Head No. 1	Column Head No. 2	Column Head No. 3	Column Head No. 4*
Verbal text here and here and here	123	Verbal text here	12.333
Verbal text here	4567	Verbal text here	456.7

*Footnote here

Figure Format

Figures normally do not have borders. They should appear as soon after first mention as possible. Figure titles (i.e., captions) are in "sentence" format, with only the first word and proper nouns capped, and with closing punctuation, even if it is not a complete sentence. They are normally flush left below the figure. If figures have notes, they should appear within the figure area if at all possible, not below the figure title. Leave some "air" around a figure. This makes the figure easier to view and improves the aesthetics of the document. Figures that are not full-page should be placed either at the top or the bottom of the page.

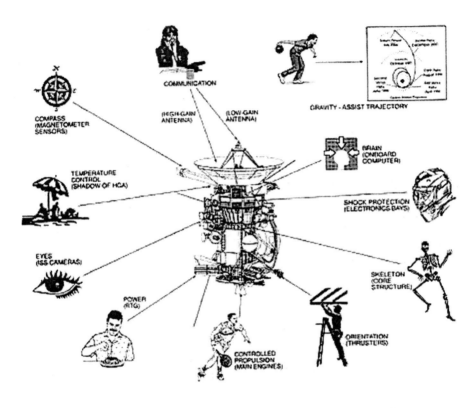

Nasa/JPL

Figure Title Here.

Warnings, Cautions, and Notes

Technical procedures often include WARNINGS, CAUTIONS, and NOTES.

WARNINGS alert the reader to situations that may be hazardous to people, and they are placed *before* the text to which they apply. They should be in boldface type. CAUTIONS call attention to situations that could be damaging to equipment, and they also come *before* the text to which they apply. They should be in boldface type. NOTES provide additional information related to processes, procedures, tooling, etc. They may come *before* or *after* the related text. Generally, the word NOTES is in boldface type, but the text of a note need not be in boldface.

There are two standard formats for WARNINGS, CAUTIONS, and NOTES, as follows:

> **WARNING**
> **In this format, the word WARNING, CAUTION, or NOTE is centered above the text, which is brought in one or twoindentions on both sides. The text is justified on the left only.**

CAUTION: **In this format, the word CAUTION (or WARNING, or NOTE) is lined up with the heading or procedural step to which it applies. The text of the W/C/N is "blocked" (i.e., has a hanging indent) on the left.**

(This is the position of the heading or step to which the W/C/N applies.)

Memory Solidifier

Instructions: Under each of the headings below, paraphrase (i.e., write in your own words) the main points under that heading in the material you have just studied. If necessary, return to the previous material to refresh your memory.

Write down, in your own words, any points related to technical writing format that are new to you or that you feel you might have trouble remembering.

-
-
-
-
-
-
-

The Writing Process

The Third Law of Good Technical Writing

Visualize your readers!

The Audience

Imagine that your readers are named JOE and MARY. Here are some questions to ask about them:

- What is their experience level?
- What is their educational level?
- How familiar are they with the type of material you are writing?
- What resources do they have available to them (e.g., tools, equipment, drawings, manuals, test procedures, in-house expertise, field service support)?
- Is English their native language?
- Are they familiar with your company's products and services?
- Does your company have a documentation or maintenance "philosophy" (or contractual requirement) vis-à-vis Joe and Mary?

The answers to the above questions will help you determine several extremely important things:

- The level of technical detail to include (i.e., what you need to *put into* your document).

- The assumptions you can make (i.e., what you can safely *leave out of* your document).
- The style of language you should use (particularly vocabulary and technical terminology).
- The types and number of illustrations to be provided.
- The references, definitions, and supporting material to be included.

In short, your goal is not to tell someone how *you* would do something, but rather to tell them how *they* can accomplish what you want them to do.

The Fourth Law of Good Technical Writing

Organize, write, read, revise . . . then do it again!

Writing as a Process

Writing is a very complex mental activity. While you are writing, your mind skips back and forth among several functions:

- *Organizing.* This involves gathering resources, organizing your environment, and coming up with a structure for your ideas.
- *Creating.* From your store of knowledge and your abilities to process that knowledge, you create something new and put it into tangible form—a document.
- *Reading.* Periodically, you reread what you have written in order to refresh your thoughts, check your focus, and plan your next step.
- *Revising.* Inevitably, you will add to, delete, or rewrite portions of your writing to improve the communication and to make sure that "all of your horses are pulling in the same direction."

Let's look at these functions more closely . . .

Organizing

Writing is time-consuming, and for most people it is hard work. So, be easy on yourself. Grab a cup of coffee, a pencil and tablet, and *plan ahead* . . . If you are not doing a standardized document, do an outline—always! Never begin writing without first doing, or studying, an outline of the document you want to create. Why? Because you are going to "organize" your piece of writing anyway--why not plan ahead instead of doing it "on the fly." Your result will ALWAYS be better.

- Organize your *content materials*. When you begin writing, you will want to have your resources available so that you will not have to stop to search for information.
- Organize your *writing space and materials*. Do everything you can to eliminate possible interruptions or distractions.

Creating

This is the act of writing. To repeat, do yourself a big favor—don't start writing until you have organized your thoughts, content materials, and writing environment. Then, go for it!

Concentrate fully on your task. The more focused you are, the better your first draft will be, the less time you will spend rewriting, and the sooner you will be able to go play golf.

Keep your momentum going. If you get stuck or do not know what comes next, reread what you have written, do some research, or study your outline again. But stick to the task. You can also skip over the sections you cannot complete and go on to other sections on which you can make some progress. Simply leave blanks to be filled in later (but make sure that you *do* fill them in later).

After you have been working and concentrating for some time, you may run out of ideas or become mentally exhausted. That is a good time to quit for a while and put the writing project aside. You can, however, let any lingering problems of content or expression "simmer on the back burner" even when you are not actively writing.

Reading

Judiciously reread your material as you are writing. It will help you to stay on track, and it will often be a catalyst for the creative process. However, do not let yourself become mired down in revising and rewriting at this time. Let your rough draft be "rough," as long as you are getting your ideas down on paper.

Revising

Most people think that revising is an agonizing chore. To professional writers, however, it is just the opposite. They know that the hard work is in *creating*. That's when you have to "dig the ditch," so to speak. So, take heart. When you get to the point of revising, the hard work has already been done!

Yet, *do not fail to revise!* A draft only becomes "writing" after it has been revised and rewritten. And for most writers, it only becomes "good writing" after the second or third read-through and revision.

Finally, if you have the time, put your document aside overnight or even for a few days. When you come back to it, you will see it with new eyes.

Some thoughts on revising:

Don't assume that your technical writers will "inspect in" quality at the end of the documentation process. They may, but *you* are the technical expert, and your name will go on the cover of the document. *Build quality in*, as W. Edwards Deming and others have said. Every minute you spend revising your work will *save* your company unnecessary labor costs downstream.

* * *

"The trouble with being punctual is that there's nobody there to appreciate it."
F. P. Jones

It's much the same with good technical writing. The better it is, the more invisible it becomes. It does not draw attention to itself but instead conveys information in a smooth, transparent, unobtrusive way.

Yet, the rewards are there:

- The knowledge that you are doing what you can to streamline the documentation process—which translates into $$$$$!
- The satisfaction that your documents are carrying the flag proudly for your company and its products and services.

Memory Solidifier

Instructions: Under each of the headings below, paraphrase (i.e., write in your own words) the main points under that heading in the material you have just studied. If necessary, return to the previous material to refresh your memory.

What is the Third Law of Good Technical Writing?

-

List seven questions to ask to help yourself "visualize" your audience:

-
-
-
-
-
-
-

List six things the answers to the above questions will do for you:

-
-
-
-
-
-

What is the Fourth Law of Good Technical Writing?

-

List two rewards of good technical writing:

-
-

Part 3

Business Writing

Identifying the Audience

Greetings . . . or, "Hello-o-o-o?"

In the previous section we talked about "visualizing the audience." That principle applies to all types of writing, but in business writing we mean something slightly different. In technical writing, it is important to think about the audience in order to convey complex information properly for that audience, at the right level of complexity, etc. This is very important in business writing, as well, but in business communication there are also some other factors involved that are very important.

For one thing, in business writing you will often be writing to people you know, but at various times you may have to communicate with people with whom you are acquainted at different levels of intimacy. When you are writing to a colleague with whom you are very familiar, your writing will probably be different from what it would be if you are writing to a superior or to a business associate at another company. Then, again, you may have to write for people you don't know, but whom you wish to influence in one way or another. Thus, business writing involves some things to think about that you don't usually encounter when, for instance, you are writing a test or assembly procedure.

Here are a few questions that will help you identify your audience:

- Where does this person stand in relation to me in the business hierarchy?
- Does this person know me, and, if so, at what level of familiarity?
- Do we have a good relationship, or has there been tension between us in the past?

- Is this person expecting to hear from me? If not, will my communication appear to be "in the course of normal business," or will it seem "unusual"?
- What are this person's principal concerns and responsibilities? Is this issue central to their main concerns, and, if not, should I ask him or her to delegate the problem?

Answering the above questions will help you to pin down who, exactly, your audience is in relation to you. Of course, we need to discuss WHY you should be concerned about these things, but before we get to that let's take a moment and talk about the overall approach of this section.

The Checklist Method

If you are a pilot or know something about flying, you know that before a pilot takes off in an airplane he or she *physically* goes through a checklist to make sure that all systems are working and that every critical aspect of the airplane has been inspected for functionality and absence of damage. To communicate well, you should do exactly the same thing.

In business writing, your first checklist should be the five points listed above. You should get in the habit of using checklists. Either commit the points above to memory or print them out and put them, along with other points you have learned, into a "communications notebook." Then, keep that notebook handy on your bookshelf in your office. Don't bury it out of sight—put it where you can get to it on a moment's notice!

Why Bother?

The main reason this is all so important is that in business writing you are almost always trying to *influence* someone. Thus, the way you present yourself in writing can be as important as the information you convey in your communication. A solid argument presented improperly just will not be as *effective* in getting results as one presented with appropriate consideration for the audience.

But what, in fact, are we talking about? You are surely asking at this point how you can shape your writing to make it the most acceptable to and effective with your chosen audience. Here are some questions to ask yourself:

- *How much time will my reader have to read my communication?* If you are writing to a busy executive, whose time for correspondence is severely constrained by other duties, be as brief as possible. One good technique is the method used by people in Hollywood when they "pitch" a story idea for a movie or documentary to a producer. They begin by presenting the very basic idea, just the kernel, and then they wait to see if the producer wants to hear more. If so, they give slightly more detail and then again wait to see if the idea "hits home." At any point, the listener can save time by saying "We've tried that idea, and it didn't work," or simply "We're not interested." Taking that technique as a model for writing to busy executives, be brief in your initial communication, presenting only the bare bones of your ideas, then close by letting them know that if they are interested and want to hear more details you can quickly oblige.
- *How well do I know the reader?* This is extremely important to help you get the right "tone" in your communication. It is probably always better to err on the side of formality, politeness, and straightforwardness (i.e., lack of irony), but on the other hand if you do have a familiar relationship with your reader it might seem "standoffish" not to reflect that personal closeness in your communication. Remember that a small bit of whimsy or irony in business communication goes a long way, but don't fail to drop your formality for a moment if you are writing to someone with whom you share that sort of relationship.
- *Do I have "good news" or "bad news" to convey?* Depending upon the answer to that question, your approach to communication may be different. If it is good news, you may want to lead off with the "news" and elaborate further down in your letter or memo. If it is not good news, you may wish to send a brief, preliminary communication alerting your correspondent to the fact that there is an unpleasant issue that must be dealt with and asking them how they want to address it (i.e., through further correspondence, a report, or a face-to-face meeting.
- *What am I trying to accomplish?* Or, alternatively: How am I trying to influence the reader? Business communication is more complex than technical writing in this sense. In tech writing, you are trying to help others carry out specific actions in the way you want them done. In business writing, your goal could be to influence

the reader in any number of ways. If you are writing a proposal, you want to cause the reader to think your company can provide the highest-quality products or services at the best price. If you are presenting your ideas on opening up a new market area, you want to influence the reader to accept your ideas. And, underlying all of your business communication you probably want to show colleagues and superiors that you are knowledgeable and articulate. Later, when we talk about specific types of business communication, we will discuss how you can shape your writing to achieve these sorts of goals.

We will elaborate on most of these topics later. For now, go to the next page and write down the main points you learned in this section. They are your first two checklists!

Memory Solidifier

Instructions: Under each of the headings below, paraphrase (i.e., write in your own words) the main points under that heading in the material you have just studied. If necessary, return to the previous material to refresh your memory.

List five points to help identify the audience:

-
-
-
-
-

List four questions to ask to achieve maximum impact:

-
-
-
-

Business Writing "Tone"

Defining Tone

Tone is a sort of "squishy" concept, but that does not make it unimportant. To start getting a handle on it, consider what Brusaw, et al. say about tone:

> "In writing, tone is the writer's attitude toward the subject and his or her readers." (p. 679)

In the previous section, we talked about identifying the audience, but the quote above introduces another idea, as well—your own feelings about the subject of your communication. By incorporating these two things into your writing, you will communicate appropriately and also with enthusiasm.

Rather than talk in generalities, let's cut right to the chase and list some things you can do to achieve the right tone. Basically, there are two broad categories of business communication, formal and less formal (or casual). As noted earlier, there can be many gradations within these two categories, but in discussing tone it helps to make this initial distinction.

Formal Tone

Here are some things you can do to achieve formal tone:

- If you are using a greeting (e.g., Dear Mr. Smith), follow the greeting with a colon. This one simple mark of punctuation places your communication on a formal, business-like level.

- Do not take too long to get to the main point of the communication. Preferably, it should be in the first or second sentence of the first paragraph.
- State things in objective terms, rather than personal terms. Instead of saying "I feel that . . ." say something like "The evidence suggests that . . . ."
- Do not use colloquialisms (i.e., slang), and particularly avoid hackneyed phrases such as the one used above—"cut to the chase."
- Begin with short paragraphs, and "cue" the reader about what is to follow by saying something like "Below I present the information we discovered on the hardware malfunction and our suggestions as to how it might be solved." This will allow the reader to assess whether he or she is the right person to receive the communication and to reroute it if necessary. It also gives the reader the option of skipping right to the conclusions.
- Use short, uncomplicated sentences in your introductory remarks. More complex sentences are acceptable, and may indeed be necessary, during your description or analysis sections. Use short sentences again in presenting your conclusions.
- Be extremely polite. If the communication has been requested, thank the reader at the beginning for requesting it. At the end, be cordial but business-like, and offer to provide more information if necessary.
- Be warm, in a business-like way. Being formal does not mean you cannot also express personal warmth. For example, if you have met your reader's spouse or know one of his or her colleagues, a short, cordial statement such as "Give my regards to Helen" or "Please give my greetings to Wesley" can add a note of warmth and friendliness.
- Keep your communication as brief as possible. If extensive elaboration is necessary, you might provide this information as an enclosure, allowing the reader to avoid reading it entirely or to assign it to someone else.
- Provide full contact information, including phone, fax, and e-mail (this information may already be on your letterhead). If you are writing to someone in another time zone, you may want to give your working hours, in case your correspondent wants to call you.

- In your closing, use "Sincerely" or "Respectfully," but not "Yours truly."

Informal Tone

Informal business writing can be a slippery slope leading to inappropriate and ultimately ineffective communication. On the other hand, studies of organizational communication have shown that a significant amount of the information passed within organizations (perhaps as much as 50 percent!) flows through informal channels. It is easy to see why. Teamwork and interpersonal support are two of the most powerful and effective mechanisms to get things done within any organization. In fact, informal business communication often grows out of successful working relationships built up over many months or years.

Here are some things you can do to give your writing appropriate informal tone:

- After your greeting, use a comma, rather than a colon. Just as the colon immediately places your communication on a formal level, so the comma sends a signal of informality and closeness.
- If possible, begin with a friendly, but brief, reference to something you and the reader have in common. For example: "Did you see the Rangers game last night?"
- Mention communications you have had with the reader in the past, perhaps on the same topic as the one you are going to address.
- Don't be afraid to express things in personal terms. For instance, you might say "As you know, I have long felt that our approach is too conservative."
- Solicit the reader's advice, feedback, or opinions. Nothing expresses mutual respect better than letting another person know that you value their expertise, thoughts, and feelings.
- Depending upon the closeness of your relationship, consider ending with something humorous. This should be handled delicately, however. If Groundhog Day is approaching, you might say "Have a nice Groundhog Day." The goal is to add warmth without seeming to be an aspiring stand-up comic.
- Close with something like "Yours truly" or "Best wishes."

Expressing Enthusiasm

When you communicate in a business environment, it is always with the idea in mind of achieving something. If you are a manager, you may be just giving orders that you want to see carried out. However, most business communication involves back-and-forth communication leading, hopefully, to some sort of action. In writing, you would like your point of view to prevail, your suggestions to be accepted and acted upon.

Thus, when you write, you should give some thought to how you feel about your subject and any collateral issues involved. For instance, if there is a design problem you are addressing, you may think that you know how to fix the problem, but you may also have some ill feelings about work done in the past or by other members of the organization. You should acknowledge these feelings, and there may even be occasions when you will want to talk about past errors or misguided policies. However, this should be done cautiously, sparingly, and with a light touch, if it is done at all. The most effective way to have your ideas accepted is to convey *positive enthusiasm*.

The second important point is that to achieve what you desire through your business communication you need to show that you *really care* about your ideas. A little passion in your writing can wake up a reader and cause that person to pay closer attention and give your thoughts more careful consideration.

Here are some things you can do to convey *positive enthusiasm*:

- *State your feelings directly.* Though this should be done with sensitivity, there is nothing wrong with saying something like "I feel very strongly about this" or "As the cognizant engineer on this project, I feel extremely confident that the new design will solve the problem."
- *Use emotional words.* Although we stated earlier that "feelings" should be left out of formal business communication, there are ways to inject your enthusiasm into even the most objective communication. For example, you can qualify your statement in such a way that the reader understands that *you* realize you are stating a feeling or opinion. You can simply say "Speaking personally, I am very enthusiastic about the new project direction." Some other emotional words that you can use are *proud, pleased, inspired, upbeat, gratified,* and *delighted.*

- *Tell the reader how you feel you can solve their problem, lighten their load, make them look good, etc.* We talked earlier about trying to identify the reader's principal concerns and responsibilities. Everyone in business has problems they are trying to solve—or have someone solve for them. If you approach someone from the point of view that you understand and have thought about *their* problems and would like to suggest some solutions, your communication will be read every time. In fact, you may even find that there is such a thing as a free lunch!

The ideas mentioned above all come down to this: if you want to communicate effectively in business, communicate at the appropriate level of formality and express positive enthusiasm.

Conveying Negative Information

Unfortunately, there are times when you simply cannot avoid communicating negative information. It's just an inescapable fact that the road of life is full of unforeseen speed bumps and potholes. And when your department or organization has hit one, the first step in correcting things is to recognize and face up to the situation as it really is.

At the same time, individuals and organizations have a vested interest in maintaining a positive self-image, which means avoiding "black eyes" career-wise or in the business community. This is not just a statement about careerism. Individuals and groups perform better when they feel good about themselves. When criticism is too harsh, morale, and ultimately quality, can suffer.

This leads to two ideas that must be kept in mind when dealing with negative information:

- "Stuff happens!" Every worker and every organization strives for perfection, yet the expectation of 100 percent perfection is overly optimistic and therefore unrealistic.
- Treat negative information like the proverbial hot potato. In fact, treat it like nitroglycerin—HANDLE WITH CARE!

The first step in communicating negative information effectively is to achieve the right tone. And because of the potentially destructive nature of negative information, the tone of the communication can make all the difference when it comes to dealing effectively with problems. Here are some things you can do to achieve the appropriate tone when communicating negative information:

- *Start with something positive.* Every employee and every organization does more things right than wrong. You can never remind yourself, your colleagues, or (even more important) your superiors too many times of how much is going right in the organization. So, find something positive, hopefully related to the problem, to use as an opener. It will give your reader the feeling that you see the big picture and that it is a basically positive picture. Then, when you present the problem, the reader will understand that you see the negative situation as a speed bump or pothole rather than a brick wall.
- *Be brutally honest—with yourself!* Since you are the one who recognizes the problem and feels compelled to communicate it to others, it is very probable that you are emotionally involved in the situation. It is quite likely that you feel frustrated in your work or disappointed that things are not going better. Thus, before you write, give yourself five minutes (time yourself!) to just sit and clarify your thoughts and to separate your emotions from the problem itself.
- *When you write, adopt a formal business tone, be as objective as possible, and under all circumstances be extremely polite.* Whether your communication is upward, downward, or in a horizontal direction within the organization, the conveying of negative information requires more respect for the reader than any other type of communication.
- *When you have written, ask yourself if your tone is such that it will effectively communicate the problem without damaging business relationships that are vital to you or the organization.* The point is that to get your reader to respond positively to your communication you must allow the reader to "save face" if at all possible. One way to achieve this is to stress that the communication is confidential. Another is to make sure, once again, that the reader understands that you see the big picture and that you are not blowing the problem out of proportion.
- *Have someone review your communication before it is sent.* Unless the communication is absolutely confidential, ask someone outside the

situation, or at least someone who is not so emotionally involved in it as you, to read your writing. Ask them specifically to give you feedback on your tone. If the communication is confidential, then your only recourse may be to play that role yourself. A good way to do that is to put the writing aside for as long as possible and then read it again when you have, hopefully, gained some distance from it.

- *Close with something positive.* To make sure the reader sees you as an ally rather than the enemy, never fail to end on a positive note. Either reiterate the positive information you used to open your communication, or find something different but related to use as your closing statement.

The decision to communicate negative information is always a tough one. Yet, organizations do not make appropriate course corrections if negative information is "swept under the rug," so to speak. Also, one of the best things you can do for yourself career-wise is to develop a reputation as one who can effectively and delicately communicate negative information when necessary. When things are going well and everyone is happy, it is not difficult to maintain positive business relationships. When problems arise, however, people immediately feel threatened, and they look for support from others. If you can get a reputation as one who is not afraid to convey negative information but who can do it appropriately and effectively—in other words, a reputation as a problem solver rather than just a griper—you will endear yourself to your colleagues, to those you may supervise, and to those to whom you report. In short, there may be no better way to put your career on the fast track!

Memory Solidifier

Instructions: Under each of the headings below, paraphrase (i.e., write in your own words) the main points under that heading in the material you have just studied. If necessary, return to the previous material to refresh your memory.

Give a definition of "tone" in writing:

-

Things to do to achieve formal tone:

-
-
-
-
-
-
-
-
-
-
-

Things to do to achieve informal tone:

-
-
-
-
-
-
-

Things to do to express positive enthusiasm:

-
-
-

Two ideas concerning negative information:

-
-

Things to do when communicating negative information:

-
-
-
-
-
-

The Importance of Brevity

It has been said that the electrostatic office copier is not the great labor-saving device it was once thought to be. The benefit of copiers, of course, is that they make it possible for more people to receive more information quickly and easily. But therein lies the danger. When the copier became a standard piece of office equipment, everyone tended to want copies of more documents. At the same time, those making copies tended to make more copies than in the past, for fear that someone would be made to feel that they were "out of the loop."

Fortunately or unfortunately, this is now an old problem, and most companies are dealing with it in one way or another. The more current problem is the computerized workstation, which now usually includes not only a computer but also a printer and a network hook-up that provides e-mail capability. Never before has it been so easy for employees to draft good-looking memos (for reproducing on the office copier) and electronic communications that can zip a message instantaneously to as many correspondents as desired. The potential for information overload is obvious.

Here's an example, a true story. One busy executive stated that when he came back from a two-week vacation he found 1,400 e-mail messages in his "in" box! It took him a day and a half to get through them all. Needless to say, he soon recommended a policy change to provide guidelines for use of the e-mail system.

This may be an extreme example, but it is one to which we can all relate today. In short, it makes brevity in communication—always valuable—more important than ever before.

Here are some ways to achieve brevity:

- Ask yourself how much information your correspondent truly needs, at this moment, to make an appropriate decision or be adequately informed about the subject of the communication. Provide enough information, but not more.
- Ask yourself how much time your reader will probably have to devote to your communication.
- Organize your thoughts—and perhaps even do an outline—before you begin writing. Pre-planning will often reveal the essence of the issue or its primary focus, allowing you to leave out things that are peripheral and not in need of immediate communication.
- After drafting your communication, but before sending it or reproducing it, revise it at least once. And think brief!
- When appropriate, use the Hollywood story-pitching technique. Provide just the kernel of your idea, then wait for feedback, perhaps asking "Do you want to hear more?" or "Would you like more detail?" If you receive a positive response, don't automatically assume that you have a mandate to send a 40-page report. Instead, provide more information, but not too much, and ask for more feedback. This technique, which admittedly may not be appropriate for all situations, can save your correspondent time and allow him or her to deal with your issue only up to the point where it may need to be delegated or redirected.

Being a good business communicator today requires that we not waste other people's time, despite the convenience of our modern means of communication. It comes down to one word—discipline. If, when you write, you "think brief," revise, and ask yourself the questions listed above, you will gain a reputation as one who can communicate crisply and effectively without putting more demands than necessary on the busy schedules of your colleagues.

Memory Solidifier

Instructions: Under each of the headings below, paraphrase (i.e., write in your own words) the main points under that heading in the material you have just studied. If necessary, return to the previous material to refresh your memory.

Things to do to achieve brevity in communication:

-
-
-
-
-
-
-
-

E-Mail Correspondence and Interoffice Memorandums

E-Mail

Although we talked about e-mail messages with regard to brevity in communication, there is more to know about using e-mail effectively. Yet, before we get to that information, there is one more thing to consider about e-mail that is directly related to message length. Let's call it the "vision factor."

What we mean by this is that people find it much more difficult to read large blocks of text from a computer screen than from a printed page (assuming that the hard copy is not a washed-out or blurred copier reprint). It is simply a fact the resolution of the letters on a screen is not as sharp as that of letters printed on a page. Of course, we are all becoming more comfortable reading from the screen, but if you send e-mail messages that are too long you risk having your reader become frustrated and perhaps irritated without really knowing why. So, the first rule of good e-mail practice is:

- Spare your readers eyestrain! Use short paragraphs . . . and send short messages.

Here are some other things to keep in mind with regard to e-mail:

- *E-mail is an informal medium.* Remember this for your own sake, if for no other reason. The length constraints (see rule #1 above) make e-mail an inappropriate medium for discussing complex or critical business issues. Even more important is that writers of e-mail messages almost invariably write quickly and off-handedly, perhaps because of the ease-of-use and the instantaneous quality of electronic mail. Ask yourself how often you have used the sort of thoughtful discipline and revision techniques in e-mail correspondence that you probably use when creating other types of written communications. If you are like most people, the answer is "seldom." Thus, think twice or three times before sending complicated or crucially important information via e-mail. The possibility of making a misstatement is very great.

- *E-mail messages, however informal, must be considered "documents of record."* Numerous court cases in recent years have established that internal e-mail messages can be used as evidence in legal proceedings. This is just one more reason to keep it simple, straightforward, and short.
- *E-mail is still a form of business communication.* By this we mean that the same courtesy and respect for the reader that you employ in more formal communications should be part of your e-mail correspondence, as well. Here are some "niceties" that can give your e-mail messages polish and also convey your respect for the recipient:
- *Always use a letter-type greeting.* The greeting need not always be as formal as in a business letter (e.g., Dear Nancy:), but nothing pleases people more than being addressed by name, whether verbally or in writing. To reduce the formality of a greeting, you might just use the recipient's first name, as follows:

Nancy,

I read your memo with great interest . . .

In any case, do not just begin talking in an e-mail. This makes people feel that you see them as too unimportant to merit a polite greeting.

- *When replying to an e-mail, do not simply tack your response onto the top of your correspondent's e-mail message to you, unless there is a very good reason for doing so.* Granted, it is convenient for jogging the recipient's mind about what they wrote and the issues to which you are replying, but receiving one's own words back is somewhat like receiving a business letter back with a few words scrawled across the top or bottom. In other words, it can make your recipient feel as though you have not taken their correspondence seriously—or, even worse, that you do not take *them* seriously enough to start a new message. The best and most elegant course is to use the time-honored methods of employing the subject line and the first sentence or two of the message to reference previous correspondence. The only exception to this is when e-mail is being used primarily to transfer a file from one person to another, in which case the attachment is very likely the "content" the recipient is seeking.

- *If you must discuss complex or crucial matters in an e-mail message, cue the reader* about the importance of the communication and urge her or him to print or download the message for their own files. To see the necessity of this, envision your message going to a busy associate and winding up in an "in" box filled with dozens of other messages competing for attention.
- *Always be courteous.* However informal your tone in an e-mail message, never forget that it is a business communication that can either reinforce or undercut a business relationship. All of the things you learned about appropriate "tone" should be brought into play in e-mail as well.
- *End with a letter-type closing.* Again, this need not be as formal as in a business letter. But e-mail is, indeed, a form of mail and thus requires a courteous closing. Here are two that work well:

Thanks,

Harry

Best wishes,

Carolyn Helms

Interoffice Memorandums

As the name implies, the interoffice memorandum is an internal form of communication within an organization. Memos differ from e-mail correspondence in three important ways. First, memos are almost always sent in hard copy, whether or not an electronic copy is sent as well. This allows the recipient to file the copy if desired, and it also reflects the formality of a memo, which is its second difference from e-mail. The third difference is that the memo form allows the writer to present a topic at some length and in some measure of detail. Interestingly enough, however, readers also expect memos to be crisp and concise, unless they have requested a lengthy report in memo format—so brevity is still the watchword.

Memo Format

Although the precise page layout of interoffice memorandums varies from one organization to another, the general framework is fairly standardized. Here is an example:

Interoffice Memorandum

February 9, 1997

To: Allison Richards, Sr. Vice President

From: William Feingold, Marketing Manager

Subject: Trade Show Booth Contract

We have now received proposals from three outside contractors for the design and construction . . .

Let's look at this format more closely. First, the heading at the top identifies the communication as a memorandum, which tells the reader that this is a formal message of some importance—in other words, it requires a degree of attention. The "To:" and "From:" lines in the example provide not only names but also titles of the sender and receiver. This is a judgement call, which must be based on the familiarity of the relationship between the two parties. However, including the titles adds a note of additional formality to the memo, and in a case such as this, where the communication is upward, using titles shows respect for one's superiors.

Finally, look at the "Subject:" line. Note that some organizations use the designation "Re:", which means "in reference to," rather than "Subject:". If this is the case in your organization, you should probably follow company practice. However, if you have discretion, use "Subject:" (and spell it out), because it is less impersonal while still being appropriately formal.

The subject line merits serious consideration for another reason. In essence, this line serves the same function as the title in a report or a chapter heading in a longer document such as a proposal. In other words, it gives the reader the first clue as to what the communication is about. In a memo, you have a bit more flexibility with regard to the subject line than you have in the title of a report. You can define your subject using a full line, or, at most, two lines of text, whereas a report title is usually much more truncated and general. In crafting your subject line, keep two things in mind:

- The subject line helps the reader to immediately determine if they are the right person to receive the communication.
- The subject line will be used to file the memo, which will in turn determine how easily it can be relocated later.

So, think seriously about your subject line. Make it direct and as short as possible, but also make sure that it expresses—and encompasses—the primary content of the memo. In the example above, a subject line such as "Trade Show" or even "Trade Show Booth" would be inadequate. The line "Trade Show Booth Contract" precisely defines the content of the memo, which allows the reader to immediately focus on the issue at hand and, even more importantly, to put out of mind any other hanging issues related to trade shows, etc.

One more thing about the subject line. Since it does serve as a "title," capitalize all words in the line except for articles, conjunctions, and prepositions of less than four letters, unless they are the first or last word in the line.

Readability Issues

Since the interoffice memorandum is a potentially complex, and therefore potentially lengthy, form of communication, there are some readability issues that should be considered. They are as follows:

- *Use 12-point type.* Whether using a serif typeface such as Times New Roman or a sans serif typeface such as Helvetica, make sure that your type is easily readable. Some faces, such as Arial, are quite readable at 10 points but under no circumstances use type smaller

than that. Also remember that serif typefaces are easier to read in hard copy that are sans serif ones.
- *Use a dignified typeface.* Today we have available to us many interesting typefaces. Yet, remember that the font you choose (i.e., both the typeface and the type size) conveys a message, however subliminal. The guiding rule should be to choose a typeface that is easy to read but does not draw attention to itself. In other words, the typeface should convey the content without getting in the way of the message. Save the fancy or unusual typefaces for posters and marketing brochures.
- *Do not right-justify.* At a distance of three feet, a fully justified page looks great. However, for best readability, use the "ragged right" format, which creates equal spaces between all words on all lines and thus relieves the reader's eyes from having to adjust continually to varying spaces between words from one line to another.
- *Use lists and headings.* A three-page memo, single-spaced, can be intimidating and hard to understand in its entirety. Because people expect to be able to get the information from a memo relatively quickly, it is a good practice to use lists whenever possible (either bulleted or numbered) to organize and draw attention to your main points. Headings in boldface are another excellent way break up the text and provide cues for the reader who may be most interested in one particular part of the discussion. Headings and lists give organizational structure to your content, which aids in comprehension and retention.

Memo Content Development

As we begin talking about the content of memos, let's reiterate two important characteristics of memorandums:

- Memos can, and often do, deal with complex and detailed subject matter.
- Readers expect to be able to extract the information from a memo quickly and easily.

These two points are related in the sense that the more complex your subject matter the more you have to be concerned about making your content clear and well-organized. Here are three other preliminary ideas to keep in mind:

- A memo should deal with only one main topic. If you have more than one topic to discuss, it is best to write a second memo. This keeps your topics separated in the reader's mind, and it makes it easier for the recipient to file the memo if desired.
- State your main topic at the beginning, in the first or second sentence. Whether or not your recipient is expecting the memo or is familiar with the topic under discussion, stating the purpose of the communication at the beginning not only lets the reader know immediately whether they are the appropriate person to deal with the issue but also orients their mind toward the subject matter.
- If you think the recipient may not want to read your memo in its entirety, refer him or her to the section in which they are likely to be most interested, or to your conclusions. In most cases, when you write a memo, you expect (or at least hope!) that it will be read from beginning to end. Referring the reader to a specific section or to your conclusions, however, can indicate that you respect the value of your recipient's time—and it may even result in a more thorough and attentive reading. In any case, it will express good will.

Methods of Development

So, how do you make your memo both detailed and easy to comprehend? One of the best ways to begin, and the one you should always use if you are writing a long or substantive memo, is to do an outline. But before you get to the outline there is an important decision that you must make. You must choose a method of development for your content. Your outline will then be dependent upon that choice. Here are the most widely used methods of development:

- *Sequential.* Use this method when writing a memo such as a trip report in which the sequence of events is important. It allows you to indicate how one event followed from another and perhaps impacted subsequent events.
- *Chronological.* In some cases, the time element is crucial, as when different but related events took place simultaneously or the

timing of the events was critical. This may or may not be a strictly sequential method of development.
- *Spatial.* When you are describing a new mechanical design, for example, you may wish to discuss different elements as they relate to one another spatially. In this case, you might also want to include an illustration.
- *Cause and effect.* When discussing a problem that has arisen, a design flaw that has been discovered, etc., you may want to examine the possible causes of the difficulty in order to provide background for suggesting solutions.
- *Comparison.* A good way to give the reader a quick, general understanding of new material is to compare it, either point by point or in broad outline, with material with which he or she is already familiar. You can also use this method to indicate that the approach used to solve an earlier problem may serve in the current situation, as well.
- *General-to-specific.* This is an excellent method to use when you have to present a detailed discussion. You begin by presenting the broad picture, then move in stages to the details. This assures that the reader will have the general background in mind against which to place the details you present later.
- *Specific-to-general.* In some cases, particularly when the reader already has a good understanding of the topic under discussion, you may wish to start immediately with specifics. Following that, you may discuss the impacts of the details on the larger issue of which they are a part.
- *Order of significance.* In some cases, such as when there are several contributing factors or pieces of evidence related to an issue and not all of them are of equal importance, you may wish to start with the most (or least) important factor and move to the least (or most) important ones.

These are, admittedly, bare-bones descriptions of the various methods of development. On the other hand, they are not difficult to understand or achieve. The important thing is to make a decision, before you write, as to how you are going to develop your material. It is also worth noting that *you may use more than one method of development in any lengthy piece of writing.*

One more thought about methods of development. Some writers—though very few—can sit down and just begin writing, organizing their material as they go. The reason is that the methods of development arise out of the ways in which we comprehend, react to, and try to manipulate the world around us. However, to be a truly effective writer, you must *bring your organizing principle to consciousness.* By that we mean that you must organize your thoughts in a straightforward, self-aware manner. One of the best ways to do this is to start your outline with a short statement of your main topic, followed by a statement of which method of development you are going to use. Let's put those two ideas in concrete:

- Start your outline with a short statement of your main topic.
- Next, put down the method (or methods) of development you plan to use in your writing.

Outlining

Finally, we have come to the topic of outlining. It is not a complicated topic, but it is very important to good writing. To outline, you simply put down your main points, then fill in the sub-topics to the level of detail that allows you to feel comfortable that you are ready to begin. Here are two things that an outline will do for you as a writer:

- It will help you to determine if you have included all of the relevant points in your argument or discussion.
- It will help you separate out peripheral material that does not need to be included or that distracts from the main topic.

Here are two additional thoughts on outlining:

- *Do it!* The great concert pianist Arttur Rubinstein once said: "Scales are hateful, but you must do them." Just so, you must discipline yourself to establish a habit of outlining. Preparing an outline not only helps you to organize your thoughts for the writing project at hand but also trains you to think critically about the structure of communication in general.
- *Computers and software do not write—people do.* There is no reason why you cannot do your outline on the computer, as long as you remember that the only "computer" that can write is the one in your head. If you find yourself skipping the outline stage in your

rush to start writing, turn off the computer and get out a pencil and paper. Once you have your outline in front of you, your internal computer will be in charge, and your writing will go smoothly and probably much more quickly.

This material on methods of development and outlining will be referred to in later sections. For easy reference, both while reading this book and later, you may wish to mark this section with a paper clip.

Memory Solidifier

Instructions: Under each of the headings below, paraphrase (i.e., write in your own words) the main points under that heading in the material you have just studied. If necessary, return to the previous material to refresh your memory.

What do we mean by the "vision factor" in e-mail correspondence?

-

List eight points to remember about e-mail:

-
-
-
-
-
-
-
-

List three differences between e-mail and interoffice memos:

-
-
-

List two points to remember about the subject line in memos:

-
-

List four things to remember about readability:

-
-
-
-

What are two important characteristics of memorandums:

-
-

List three ideas to keep in mind regarding memorandums:

-
-
-

List the most commonly used methods of development:

-
-
-
-
-
-
-
-
-

List two things to remember about outlining:

-
-

What two things will outlining do for you?

-
-

List two final thoughts regarding outlining:

-
-

The Formal Business Letter

Tone

Fifty years ago, business letters of all types were much more formal and objective in tone than they are today. In fact, many textbooks written over the past few decades have counseled writers to loosen up on the formality in business correspondence. Yet, with the advent of e-mail, facsimile transmission, etc. in today's business environment it may be time to remind ourselves that a business letter, whatever its purpose, is still a formal type of communication. To refresh your thoughts on this issue, you may want to reread the section on business writing tone.

Letter Format

Date of Correspondence

In a formal business letter of any type it is extremely important to include the date of correspondence at the very top, either flush left or right. Business letters are often filed, and while the date of receipt is sometimes stamped on them this is not always the case. The date on a letter is often vital information for those receiving or processing the letter, and it also allows the recipient to refer to the letter by date when responding. If several letters on the same topic are passing back and forth between correspondents, the date of each letter becomes even more important.

Contact Information

In this age of electronic communication, it is amazing how often writers forget to include full contact information in their correspondence. Letters written on company letterhead stationery include information on how to contact the company, but not necessarily the person writing the letter. And letters written on plain bond paper, such as job application letters, contain no contact information at all unless the writer includes it. Thus, it is important to think clearly about whether, how quickly, and by what means the recipient may wish to contact you after receiving your letter.

Traditionally, letters written on non-letterhead paper included a full return address immediately under the date and above the internal address (the address of the recipient). This practice has loosened up considerably today because of electronic communication, but the function and importance of the return address have not gone away. Today, it is generally acceptable to place contact information below the signature line at the bottom of the letter. The amount of information included may vary depending upon the purpose and recipient of the letter, but it can include a full return mailing address, phone number(s) with a notation as to the availability of message recording service, a fax number, and an e-mail address.

Note that the contact information should be placed either at the top or the bottom of the letter, rather than being buried in the body of the letter itself. The reason for this is that the recipient, after receiving the letter, will often file it and come back to it later, when he or she wishes to contact the writer. Putting the contact information in a clearly visible place on the page makes it easier for the respondent to locate it at a glance.

The Internal Address

It is still standard format for all types of business letters to include a full internal address (including zip code), positioned flush left. It may seem old-fashioned, but the internal address is a formal expression of professional courtesy, which tells the recipient that the correspondence is on a business-like level. Today, with the explosion of form letter correspondence, such courtesy is not always observed, but correspondence between two individuals or an individual and a business organization should in all cases include the internal address.

The Greeting

As with other aspects of business letter format, the greeting has become slightly less formal over the years. There are three basic situations, as follows:

- *Writing to someone you know.* If you are writing to a business associate with whom you are well acquainted, it is acceptable to use just their first name in the greeting (e.g., Dear Marilyn:).
- *Writing to someone whose name you know but with whom you are not well acquainted.* In this case, use either "Mr." or "Ms." followed by the recipient's last name (e.g., "Dear Mr. Franks:").
- *Writing to an organization.* When the only information you have is the name of the organization or department (e.g., Human Resources Dept.), use the greeting "Dear Sir or Madam:" (traditional), or "Dear Mr. or Ms.:" (more contemporary).

Note that in each of the three cases above the greeting ended with a colon. In all business letters, except perhaps those between associates who are also good friends, the comma is inappropriate. Much like the internal address, the colon in the greeting is a mark of professionalism that shows respect for the business-related character of the communication.

The Body of the Letter

Few business correspondents indent the first lines of paragraphs anymore, although that format is still acceptable. The most common paragraph format is blocked on the left, ragged on the right, with a blank line between paragraphs.

As in all types of written communication, do not leave the first line of a paragraph hanging alone at the bottom of a page (called an "orphan") or the last line of a paragraph hanging alone at the top of a page (called a "widow"). For "orphans," move the line to the top of the next page. For "widows," move at least one more line to the second page.

Closing

In days gone by, closings were often more florid than they are today (e.g., "Anticipating your reply, I remain, . . . Yours sincerely,"). Business letter writers today invariably use shorter, less ostentatious closings. The two most common are "Sincerely" and "Respectfully." Follow the closing with a comma. If you use a closing of more than one word (e.g., Yours truly), the second and all succeeding words are not capitalized.

After the closing line, leave at least three blank lines (i.e., four returns) before typing your full name. This will give you room for a signature that does not look cramped. Traditionally, closing/signature lines were often placed at the right, lined up under the date and return address. This practice has largely disappeared, along with indented paragraphs, but it is still acceptable. The most common format is to place the closing flush left.

Additional Information

Three types of extra information may be included below the signature (with at least one and preferably two blank lines between). This information is as follows:

- *Typist's initials.* If it is company policy to include this, the format is the letter writer's initials (in caps) followed by a colon followed by the typist's initials in lower case (e.g., DRM:atg).
- *Enclosure indication.* When there are one or more enclosures with the letter, it is professional courtesy to indicate so at the bottom of the letter. This lets the recipient check to see that everything has been forwarded as intended. There are two possible formats:

Enclosures: (3)

Enclosure: Product Support Summary

- *Copy information.* This information is conveyed using the designation "cc:" (which originally stood for carbon copy) followed by the name of the person who is receiving a copy. If more than one person is receiving a copy, list the names vertically, either in alphabetical order or by order of rank within the organization. Note that the "cc:" is not repeated after the first line.

Pagination

Page one of a business letter is never paginated. On all succeeding pages (see the discussion of letter length below), place page numbers at the bottom, centered.

Letter Content

Introduction

Every business letter, whatever its purpose, begins with an introductory paragraph. Generally, this paragraph is short, but that does not mean it is unimportant. In terms of achieving your purpose, in fact, it could be the most important paragraph in the letter. The introductory paragraph is used to accomplish two things:

- Establish the appropriate type of rapport with the recipient.
- Orient the recipient's mind to the subject to be discussed.

Writing a good introductory paragraph is much like meeting someone in person in a business situation. In fact, thinking spatially is a good technique. For example, when you meet someone with whom you are very familiar, you feel comfortable moving in close and perhaps extending your hand. When you are meeting someone for the first time, you will probably stand a few feet away and wait to be introduced or introduce yourself verbally before moving closer. If there is a difference in rank, and especially if you are meeting a superior, you will undoubtedly (though perhaps unconsciously) respect the other person's physical space until invited to come closer.

Just so, in your introductory paragraph a formal tone indicates respect for the recipient's business-related "space." A less formal tone, on the other hand, indicates that you recognize the relationship you have established with the reader and appreciate it. It is rare that a "howdy!" tone is appropriate in a business letter, but it is not out of the question if handled delicately.

In some ways it is difficult to generalize about how to start a business letter, but it is safe to say that from the first word—literally—the reader is reaching not just for information but is also processing your tone to see if the rapport is appropriate.

The second thing the introductory paragraph does is quickly direct the reader's mind to the purpose of your letter. If you have to write many business letters, you should practice writing opening paragraphs. They should be no more than two sentences long, and they should accomplish the two things described above. The best way to make sure that your thinking is focused is to discipline yourself to state your purpose in one or two sentences. If it takes more than that, you probably have information to communicate that is too complex for a letter and should instead be put into a report. Let's codify that rule:

- The introductory paragraph of a business letter should be no more than two sentences long, if at all possible.

The Substantive Paragraphs

The structure and tone of the substantive paragraphs of a letter can vary significantly, depending upon the purpose of the letter. A letter of recommendation, a cover letter for a report, and a job application letter are quite different in terms of the content required. However, good organization is still paramount. For guidance on structuring your content, you should review the material on Methods of Development in the section on Interoffice Memorandums.

A business letter, however, differs considerably from a memorandum. Whereas a reader may tolerate a six-page memorandum (if well organized!), the same reader probably would not appreciate receiving a six-page letter. The reason lies mostly in people's expectations regarding different means of communication. People expect letters to be short, succinct, and right to the point. They expect them to have a well-defined business purpose (e.g., congratulations on a job well done, covering information for a business proposal, etc.), and they do not want to have to wade through information unrelated to the main purpose. Thus, here are three guidelines that should be considered 90 percent "firm."

- Overall length should be no more than 1½ pages.
- Paragraphs should be short (no more than four sentences).
- Nothing should be included that is not directly related to the main topic.

Closing Paragraph

The closing paragraph of a business letter serves much like the summary of a report—except that it has to be much shorter and must accomplish even more. In the closing of a letter you must both wrap up the substantive content and conclude what is in fact a communication between two individuals. Here are some guidelines for writing good concluding paragraphs:

- It should be no more than three sentences long.
- The first sentence (or two) should either summarize your discussion or reiterate, in a nutshell, your main point.
- The last sentence should be a cordiality that expresses interest, invites feedback or response, or indicates a willingness to provide more information or a more in-depth analysis of the topic discussed in the letter.
- Brake smoothly! End your business letters the way you brake for a traffic light when there are others in the car—confidently but comfortably.

Memory Solidifier

Instructions: Under each of the headings below, paraphrase (i.e., write in your own words) the main points under that heading in the material you have just studied. If necessary, return to the previous material to refresh your memory.

A business letter is what type of communication?

-

List two reasons the date is important:

-
-

Contact information can be placed where?

-
-

Why is the internal address important?

-

List three types of greeting situations, and give examples of appropriate greetings:

-
-
-
-

The greeting in a business letter is followed by what type of punctuation?

-

What is an "orphan"?

-

What is a "widow"?

-
-

List the two most common closings:

-
-
-

List the three types of additional information:

-

What is the rule of pagination for business letters?

-
-

What two things should the introductory paragraph accomplish?

-
-

The introductory paragraph should be how long?

-

List three guidelines for writing substantive paragraphs:

-
-
-

List four guidelines for writing good concluding paragraphs:

-
-
-
-

The Internal Report

Introduction

The technical or business report is a form of writing used to communicate detailed and often lengthy information within organizations. Because of its detailed content and formal style, the report falls somewhere between actual technical documentation and the memorandum. In some organizations, reports are not used as frequently as the more standardized forms of communication (such as engineering documentation), but in other organizations, especially those for which research is the major activity, a report may be the final outcome of many weeks or months of effort.

There are many types of reports. Indeed, the report form can be used for almost any business purpose where an extensive amount of information must be communicated in written form. For that reason, we will not take time to identify and analyze the various types of reports but will instead focus on some common themes of report writing.

Report Style

Reports are generally requested, and therefore they are expected by their recipients. Thus, it is not necessary to spend time explaining why you are communicating. The writing style should be objective in tone, with personal feelings and opinions removed except to the extent that you draw conclusions from the data you present in the body of the report.

For a sense of what report style should be, review the section on Technical Writing Style in Part 2 and the section on Business Writing

"Tone" earlier in Part 3. In general, remember these two points about report writing:

- The "tone" of a report should be formal.
- The writing style should be objective.

Report Structure

Reports may contain the following sections, whether or not they are labeled as such:

- Abstract
- Summary (sometimes called the Executive Summary)
- Introduction
- Body
- Results
- Conclusions and Recommendations
- List of Abbreviations and Acronyms
- References
- Appendixes

Now, let's look at each of sections more closely.

The Abstract

Although we will not discuss them in this book, formal scholarly papers almost always include an abstract. Reports may or may not contain an abstract. If it is required, the abstract is normally a one-paragraph summary of the report that quickly gives a glimpse of the report's purpose, methodology, findings, and conclusions.

The main purpose of an abstract is to provide a small amount of information about the report so that it can be included in a searchable database of some kind. It may also serve to give readers a brief synopsis of what they are about to read.

If an abstract is expected, spend an adequate amount of time on it, and rewrite it at least once. A good abstract may make the difference between a report being ignored and its being given widespread consideration within the author's field or industry.

The Summary

The summary (or executive summary) is most important in lengthy reports. The reason it is important is that reports are often distributed to readers of different kinds and at different levels within an organization, not all of whom are interested in the details of the study or project. For example, the horizontal distribution may include technical professionals in other disciplines who nonetheless want to be kept up to date on the progress of a project being conducted by their colleagues. The vertical distribution may include upper-level managers who want a short overview to keep them appropriately informed. In cases such as these, a well-written one- or two-page summary can be extremely helpful and welcome.

The Introduction

The introduction is extremely important, even in longer reports that include an executive summary. The reason is that readers who intend to read the entire report may skip over the summary and dive right into the report itself. The words of caution about writing a good abstract or summary also apply to the introduction, and for reports that do not include a summary the introduction becomes even more important. Here are several things that a good introduction should accomplish:

- Express the purpose of the report.
- Provide a brief justification for the report—why it was written.
- Indicate any previous research, events, or experience upon which the report builds.
- Briefly outline the methodology used to address the problem.
- Present the main results of the study, design effort, etc.

Here is a rule to follow in writing your introduction:

- Make the introduction as short and tightly written as possible without leaving out any pertinent information.

How do you decide what is pertinent information and what is not? Here are two guidelines that you can consider to be the "purposes" of the introduction.

- To focus the reader's mind on the topic to be discussed.
- To provide an organizational structure upon which the reader can "hang" the details presented in the body of the report.

The second point above is especially important. People grasp and retain information much more easily if they have a structure upon which to hang new material. This structure should include the following two things whenever possible:

- Related past experience or information with which the reader is familiar.
- The organization of the ideas—i.e., how they relate to one another.

In some cases, when you have struggled with your introduction and do not feel that you are getting it exactly right, you may wish to draft the body of the report first, then go back and write your introduction. In that way, when you approach your introduction you will know the structure and content of what is to come, because you will have already written it!

The Body

The body of a report can vary considerably, depending upon the type of report it is. One of the most important things about the report form is that it gives you the space to provide background information, theory, and a full description of the steps you followed to address the problem or issue. On the other hand, if you want your report to receive the consideration and thorough reading it deserves, it should not waste the reader's time by making them wade through material that is not directly pertinent to your topic or argument. With regard to scientific solutions, Einstein once wrote that they should be "as simple as possible, but not simpler." Let's paraphrase that as a rule for writing the body of the report:

- The body of a report should be as succinct and "lean" as possible—but not more than that.

How do you achieve that strong focus and discipline? Organization! Specifically, revisit the material you encountered earlier on these two topics:

- Outlining.
- Using an appropriate method of development.

Results

This section may not always be relevant or necessary, but if you are reporting on a research project, a problem resolution effort, etc., the results section may be the most significant for the reader. Indeed, the reader may be somewhat familiar with the ongoing project and is now anxious to find out "what happened."

The content of your results section should flow directly from the body of your report. In fact, it is a good policy to write the body with the idea in mind of bringing the reader up to the point where he or she understands what was done and now can easily grasp the results. Thus, if you write a good report body, the results section may be very short, indeed.

Conclusions and Recommendations

In some reports, the results are all that need to be reported. In other words, some studies have no policy implications but rather deal with clearly defined issues. If you are reporting on a product testing effort, the results (pass or fail) may be all that the reader wants to know. However, if you are reporting on a new marketing campaign there may be policy issues or multiple courses of action available. In this section you outline your conclusions based on your study and provide recommendations based on those conclusions. Here are three guidelines for writing a good conclusions and recommendations section:

- The conclusions should be solidly supported by the results of the study.
- Your conclusions should be identified as such and not presented as "facts" until they have been proven, replicated, etc.
- Any recommendations offered should flow directly and clearly from your conclusions.

Abbreviations and Acronyms

In long reports (say, more than ten pages), include a list of all abbreviations and acronyms used. Your principal reader(s) may not need such a list, but your distribution may include others who are either horizontally or vertically distant from the project but who still have a "need to know." Remember—all it takes is one unfamiliar acronym or abbreviation to alienate a reader. Why take the risk, especially after you have expended great effort not only in doing your study or design but also in writing it up? Here are two rules for creating a list of abbreviations and acronyms:

- Arrange them in alphabetical order by short form, *not* by the spelled-out term.
- Include every one of the abbreviations and acronyms used in the report, however familiar you may feel they are to the reader.

References

If you have cited previous studies or other informational sources in the report, you should provide a list of references that gives full publication information. This allows the reader to access these sources easily if she or he desires further background information. Here are two rules:

- *In-text citations should include the author's last name followed by the year of publication. Example: Smith (1998) or (Smith, 1998).*
- *In the reference list at the end, references should be in alphabetical order by the first author's last name. For standard bibliographical format, see a style guide such as the Chicago Manual of Style, published by the University of Chicago Press.*

Appendixes

(NOTE: Yes, the construction "appendices" is still correct. Both forms are commonly used, and the only rule is to be consistent throughout your document.)

One of the nice things about the report form is that it allows you to "layer" your communication somewhat by placing information that may be of interest to some, but not all, of your readers in one or more appendixes. How do you decide whether to include information in an appendix rather than in the body of the report? Interestingly enough, the guideline is something we talked about at the very beginning of Part II: identifying the audience. You may wish to reread that section to refresh your memory. Meantime, here is the guideline:

- *To decide whether to put material in an appendix, analyze your audience, both in depth and in breadth.* Ask yourself such questions as: What do my readers want to know right now? Do my readers vary in terms of the level of detail in which they are interested? Will it make my report more effective and focused if I move some material to an appendix?

Final Thoughts on Reports

The report is a flexible and therefore extremely versatile form of communication. But therein lies the danger. Because the report form is less structured than, say, a test procedure or product specification, it can become verbose and thus less than completely effective. Here are two rules for using the report form:

- Use all of the parts of a report to guide your reader(s) toward one compelling conclusion—yours.
- Because the format of a report is flexible, use an extra measure of discipline in organizing your thoughts and planning your method of development.

Memory Solidifier

Instructions: Under each of the headings below, paraphrase (i.e., write in your own words) the main points under that heading in the material you have just studied. If necessary, return to the previous material to refresh your memory.

List two points to remember about report style:

-
-

List the parts of a typical report:

-
-
-
-
-
-
-
-

What is the main purpose of the abstract?

-

Why is the summary section important?

-

List five things a good introduction section should accomplish:

-
-
-
-
-

What is the rule for writing a good introduction?

-

What are the two purposes of the introduction?

-
-

List two things to keep in mind in order to facilitate comprehension and retention:

-
-

What is the rule for writing the body of the report?

-

List two ways to achieve focus and discipline in a report:

-
-

The information in the results section should flow directly from where?

-

List three guidelines for writing a good conclusions and recommendations section:

-
-
-

List two rules for creating a list of abbreviations and acronyms:

-
-

List two rules for creating a reference list:

-
-

What is the rule for deciding whether or not to put material in an appendix?

-

What are the two general rules for using the report form?

-
-

The External Proposal

Introduction

One of the most important types of documents in the business world is the formal proposal. It is the instrument by which many businesses acquire new contracts or additional business from established customers. Two things should be noted as we begin our discussion.

First, while there are some occasions when proposals are written for internal consideration, particularly within very large corporations, the principles of proposal writing are the same whether the "customer" is upper management, another division within the company, or an external organization. In this section we will approach the subject from the perspective that the customer is external and that the proposal is in response to a formal Request for Proposal or RFP.

Second, in most companies the proposal process functions alongside, and in coordination with, other marketing and sales activities such as advertising, trade show marketing, etc. This is an important thing to keep in mind, because it highlights the fact that a proposal, however detailed and technical, is also a selling tool. Thus, the proposal form is a blend of technical problem solving and persuasion, and good proposal writing is as much an art as a science. On the technical side, the proposal must be rigorous, accurate, and (hopefully) innovative. On the marketing side, the proposal should convey enthusiasm and confidence without either over- or under-selling the company, its products, or its capabilities.

Let's restate those two things succinctly:

- The principles of good proposal writing apply to all types of proposals, whether internal or external.

- A high-quality proposal functions as both a technical document and a selling tool.

Proposal Content

The structure of a proposal is almost always determined, to one extent or another, by instructions in the RFP. Some RFPs allow proposers a certain amount of leeway with regard to section headings, number of sections, etc., while others provide very detailed instructions, including the exact wording of major headings, the number of pages allowed under each, and even the font (both typeface and point size) to be used. Such detailed guidelines often strike proposal writers as arbitrary "whims" on the part of the requesting organization (and they may be), but it is likely that some thought has gone into them. Requesters of proposals may, first of all, need to place a reasonable limit on the amount of material they must review at the first stage of a procurement process. Also, and probably more important, they need to level the playing field as much as possible so that all proposers are given an equitable evaluation. One of the ways of doing this is to limit all proposers to the same number of pages of text under each major heading.

On the basis of these ideas, here are two vital rules to remember when starting any proposal process:

- Assume that the requesting organization has a reason for every guideline in the RFP, even if you do not know what that reason is.
- Read the RFP carefully at the outset, highlighting all passages that deal with content limitations, formatting, etc.

The two rules above are so important that it is worthwhile elaborating the reasons why they should be followed:

- Failure to comply with content or format guidelines can cause your entire proposal to be judged "noncompliant."
- Proposal efforts are extremely labor-intensive, and thus very expensive. Having to rewrite a proposal to fit RFP guidelines can double the expense!
- *Proposals are deadline-driven.* Remember, having a cut-off date for receipt of proposals is a fairness issue on the part of the requestor. After all, you would not want other proposers to have more time

than your organization. The point? Plan ahead! In other words, know the guidelines, and work to them from day one of the proposal process. Use your precious time to create a winning proposal, not to rewrite or reformat.

With regard to the specific content of proposals, little more can be said. However, there is a theoretical framework that should be kept in mind. It is essentially this: an RFP is a formal expression of a need, problem, deficit, etc. on the part of the requestor. The job of the proposer is, first, to identify the need (which may or may not be well articulated in the RFP), and, second, to propose a solution. Anyone who has worked on a proposal understands the second of these concepts. It is the first that is often neglected. Don't do it! Instead, do this:

- Restate the requestor's problem or need in different words. This is vital. It will do two things for you:

    - It will force you to look closely at the problem as stated in the RFP. This effort may uncover hidden assumptions on the part of the requestor or "the problem behind the problem." For example, if you are bidding to provide engineering support for the construction of an oil refinery, you should consider that the customer's desire is not for an oil refinery but rather for increased profits through the addition of more refining capacity. This subtle distinction could help you to create a design for the plant that is not only technically right but also cost-effective.
    - It will help to assure that all of your subsequent efforts are tightly focused on addressing the customer's need. During a proposal-writing process, loss of focus can be extremely costly in terms of both time and money.

Now, what about the solution, itself? It is imperative to keep in mind that the requestor is not just looking for a technical solution to a problem, but rather for a technical solution at an acceptable price within a desirable time schedule. The appropriate "solution" for the customer, then, must include all of these aspects. Let's look at them more closely and see how they interact.

- *The technical solution.* Obviously, this is the most crucial aspect of the proposal process, because if you cannot solve the customer's problem you may as well not spend the time and money writing a proposal. But let's say that you are sure that you *can* solve the problem or fill the customer's need (perhaps you have done similar things many times for other clients). The immediate next step—before you even *begin* to flesh out the details of the technical section—is to integrate the potential solution with the requestor's cost and schedule requirements, if they have been stated.
- *The cost aspect.* The RFP may give some indication of the budget within which proposers must operate. If so, this will probably have a significant effect upon the technical aspect of the proposal and upon what can, in fact, be proposed. In some cases you may not feel that the cost constraints are realistic, but regardless you must either propose within the limits or provide justification as to why you feel the limits are unreasonable. Most often, however, it is assumed that the requestor is interested in the lowest possible cost that will still assure a quality solution to their problem. Thus, it is obviously in the self-interest of the proposer to keep the price as low as possible. The temptation at this end of the spectrum is to unconsciously "low-ball," or to price one's services too low, out of a desire to win the contract. The caution here is that the requestor is probably very "savvy" about the industry and what various solutions and levels of services cost. The slightest hint that you are low-balling on price can alienate the requestor's review committee from an otherwise good proposal. The lesson to be learned is that the only way to win in the proposal process over the long haul is to coordinate the technical and cost aspects accurately and realistically from the outset.
- *Schedule.* Every requestor and every proposer knows that schedule can impact both cost and "scope"—i.e., the complexity of the solution, the level of services offered, etc. It is more common for schedule to be explicitly stated in RFPs than cost constraints, and thus the caveats noted under cost above are even more important here. The keyword, again, is coordination. And the reason, from the writing perspective, is plain. Proposal writing is expensive to begin with, and rewriting a proposal at the last minute can be devastating to the proposal-writing budget.

From the above information, we can state a general rule for keeping the costs down during proposal-writing efforts and for writing winning proposals:

- From day one, coordinate the technical, cost, and schedule aspects of the proposal.

Here are some other guidelines for proposal writing:

- *Read the RFP carefully, and continue reading it.* Each member of the proposal team should have a copy of the RFP in a three-ring binder, and any updates or clarifications to the RFP received from the requestor should be given to each team member for inclusion into their binder. Within the first few days of a proposal effort, each team member should read the RFP at least twice. The reason is that as you read an RFP the first time you are probably thinking ahead to the solution you are going to propose. In this state of "divided consciousness" you are likely to gloss over or misread some aspects of the RFP. Thus, it is important to reread the RFP again after your thinking has begun to coalesce. Then, reread the RFP, or portions of it, at significant junctures throughout the proposal process.
- *Identify the audience.* Find out as much as you can about the requesting organization. Has any member of the proposal team dealt with them in the past? Do they have a Web site that you can study? Today, with so much information about organizations online, companies expect you to do some research and to know who they are, what they have done, their major product lines and customers, etc. Try to "put a face" on the customer and identify the company's goals, market position relative to competitors, and financial resources. All of this can help you write a more focused proposal that addresses both the explicit and implicit needs of the requestor.
- *Analyze your own assumptions.* For example, does your corporate culture contain an unstated assumption that "we only provide high-end solutions"? Perhaps the requestor is looking for a more modest approach, cost-wise. Does this mean that you do not want their business? Or, are you willing to propose a solution within the budget constraints of the requestor? Answers to questions such as these only come to the surface when you make the effort to identify

and analyze your own company's underlying assumptions about how to do business, what solutions are acceptable, what things cost, and how quickly things can be done. One of the best ways to accomplish this is to appoint one member of the proposal team to think precisely about these types of things—that is, to play "the devil's advocate." This allows other members of the team to think creatively without constraints but also assures that underlying assumptions will not go completely unexamined.

Boiler Plate Material

The term "boiler plate" is widely used to refer to material that is recycled in proposals and other publications. Examples of typical boiler plate are company profiles, descriptions of past projects, lists or descriptions of company resources, and resumes of managers and technical staff members. Given the often punishing deadlines under which proposals must be written, having some good boiler plate on the shelf can be an immense benefit. There are dangers in using boiler plate, however. Here are three guidelines for using boiler plate effectively:

- *Reread it before using it!* Every piece of boiler plate should be reread *word-for-word* before it is included in a new proposal. Terrible gaffes have happened because boiler plate was simply "plugged in" without review. Resumes may need to be rewritten to emphasize experience relevant to the proposal. Company resources may have changed (hopefully, for the better!), and these additions should be included. And histories of company projects are *always* out of date!
- *Format it correctly!* Today, some RFPs are very specific about formatting, including margins, type, etc. Boiler plate, like everything else in the proposal, must meet the requestor's requirements.
- *Get the boiler plate ready as early as possible!* Because it is "only boiler plate," this material is often not pulled together until the end. Then, to everyone's horror, it may be found that the resumes are too long and "bust the page-limit budget." Or—and this has happened to every company—you go looking for a piece of boiler plate at the last minute, only to find that it has disappeared, resides on a password-protected computer, or must be entirely rewritten. Ah, the human memory! How often it plays tricks on us! What we thought was there on the shelf, ready to go, is instead a new

problem to be solved in the middle of the night before the proposal needs to be shipped. How to avoid having this happen? As with identifying hidden assumptions, the best way is to assign someone early in the process to get the boiler plate together. And give them a deliverable date!

Proposal Writing as "Writing"

You may be saying to yourself right now that this section did not talk very much about writing, as such. One reason is that, if you have applied yourself during the earlier sections of this book you have picked up most of what you need to know about the act of writing. The second reason has to do with a concept that you have hopefully taken to heart by now, which is that good writing is more than just putting pen to paper. It involves planning, organization, and mental focus, all of which, to some degree at least, *precede* the actual work of sitting at the keyboard.

Yet, there are two ways in which proposal writing differs from most other types of technical and business writing. Let's look at them.

- *Proposal writing is writing under the pressure of a tight deadline.* If you do not have to work in this way on a regular basis (as, say, daily newspaper journalists do), the urge that probably comes over you at the beginning of a proposal effort is "let's get going" because "time's a-wasting." Good proposal writing requires that you resist this urge, while keeping the deadline always in mind. The irony is that *when working under a deadline, pre-planning is more important that ever*. The reason, of course, is that you don't have time to go back and fix things too many times.
- *Proposal writing is group writing.* Whenever something is being written "by committee," so to speak, coordination of the efforts of various team members is absolutely essential. Today, that takes two very important forms:

    - *Coordination of content.* In a winning proposal, all content must be focused on the same solution, and there must be as little duplication of effort as possible. Someone, perhaps the proposal team leader, needs to oversee this aspect.
    - *Coordination of computer files.* From the outset, all writers should work in the same software, and hopefully on the same

computer platforms. If files are going to be passed around for review and comment, the concept of document change control must be taken seriously. Steps should be taken to assure that in the end only the final draft, with all reviewers' comments included, is put into the proposal. In today's flexible communication world, this takes some planning and thought. If things become too messy, you can always do things the way they were done in the past—tape hard copies of each section to the wall in the proposal "war room" and let reviewers mark them up by hand. That way you can control all changes at one central location or work station.

Editing and Graphic Design

Before we finish our discussion of proposal writing, it is important to recall that proposals, generally speaking, are external documents. And, more than that, they are selling tools. In other words, they are not just going out there into the world, they are also trying to convince others that your company can provide the best solution, on time, at the best price available. Thus, you don't want them to simply be accurate. You want them to read well and look good, as well.

Let's state that another way: *"A well-written, well-designed proposal might not win a contract . . . but a poorly written, poorly designed proposal could lose it!"* You may say to yourself that the customer is "only interested in the information." Don't fool yourself! A proposal that is not well-written and nicely designed probably isn't well organized, either. It's all of a piece. Don't get stung. Here are two final guidelines for creating winning proposals:

- *Always have proposals edited by a professional technical writer or editor; have boiler plate re-edited every time it is used.* This is especially important because proposals are often written by more than one writer. Left unedited, a proposal is only as strong as the skills of the weakest writer.
- *Have a graphic designer on hand as a consultant, whether or not you use one to create an overall design for the proposal.* Graphic designers are trained to think visually, to make things not only attractive and easy to read but also emphatic. Why not have their skills in your toolbox when writing proposals?

Memory Solidifier

Instructions: Under each of the headings below, paraphrase (i.e., write in your own words) the main points under that heading in the material you have just studied. If necessary, return to the previous material to refresh your memory.

State two general ideas about proposal writing:

-
-

State two vital rules to remember when starting a proposal process:

-
-

Give three reasons *why* the above rules are important:

-
-
-

What is the first step in the proposal process?

-

What two things will this do for you?

-
-

What are the three aspects of the "solution"?

-
-
-

State a general rule of proposal writing:

-

List three other guidelines for good proposal writing:

-
-
-

List three guidelines for effective use of boiler plate material:

-
-
-

State two reasons why proposal writing is different from other types of business or technical writing:

-
-

List two types of coordination that are important in "team" writing:

-
-

State two final guidelines for proposal writing:

-
-

Presentations

The Presentation as a Piece of Writing

A presentation, whether delivered at a meeting or a luncheon, is an oral form of communication, but it is also formal. Thus, it must be considered a piece of written communication, as well. A presentation should be prepared in advance as carefully as any other type of written communication.

Let's review the principles of the writing process, since they apply to presentations as much as they do to reports, memorandums, and technical documents.

- *Define your purpose.* Before you begin to write or to create your visuals, put down the purpose of your presentation in one sentence. Because of the time constraints inherent in the presentation form, there is no room for anything that is not directly related to your main theme or purpose.
- *Identify your audience.* Your audience is not "the Quality Council" or "the Executive Council." Think of your audience as being composed of individuals, and put "faces" on them if at all possible. Ask yourself what their needs and problems are and what you can do to make *their* lives easier.
- *Create an outline.* If you think you don't have time for this step, you will probably end up wasting your audience's time in the end.
- *Develop your content.* Note the word "develop," as in "methods of development." Think as critically about the structure of your presentation as you do about the structure of anything else that you write. If necessary, review the methods of development you learned earlier.

- *Revise!* Good writers *always* revise, and because they do they build into their schedules enough *time* to revise their work.
- *Practice your delivery.* Practicing in front of a mirror is an excellent way to get a feel for how you are coming across. Giving your presentation to another person as a "dress rehearsal" is another good way to prepare. A third way is to record yourself, either on audio tape or videotape. Practicing beforehand will accomplish two very important things for you:

  - *Allow you to time your presentation.* Whether you have a time limit or not, you probably have a sense of how long your presentation is expected to be. Try to hit the bull's eye, time-wise—i.e., neither too short nor too long for the occasion and the subject.
  - *Provide you with feedback.* The more prior feedback you can get on your delivery and your content, the better your final presentation will be. Do what theatre people do: establish a rehearsal schedule, and stick to it.

Characteristics of Presentations

In no particular order, here are the principal characteristics of presentations that you should keep in mind:

- *The tone should be formal.* The attention of an audience of busy people is a valuable thing, and you should treat it as such. Under no circumstances treat your listeners as a "captive audience." One way to convey respect for your listeners' time is to use an appropriately formal tone, in both visual content and oral speech. Use humor, casualness, and other expressions of "personality" sparingly and cautiously.
- *Brevity is paramount!* The requirement of brevity is more stringent in presentations than in any other form of communication, and the reason is that the audience is, in some sense, "captive." To comprehend your argument, your listeners have to sit through your full presentation (unless they have been given a hard copy), and at the end you want them to feel that they have had to give up only a reasonable amount of their time relative to the "payoff." So, think "brief" from two perspectives:

- Keep the overall presentation as short as possible (though not shorter).
- Keep each visual as short and uncomplicated as possible. Nothing alienates listeners more quickly than the feeling that they cannot "keep up." In fact, listeners expect to be able to keep up with your presentation without taxing themselves too much. It's in the nature of presentations. So, give your audience that "warm and cozy feeling."

- *A presentation is not for detailed arguments.* This is closely aligned with the requirement for brevity. Presentations are oral communication with visual aids. Thus, they are linear, and the listeners have only one chance to grasp the content as it goes by. Even if you have handed out a hard copy of your presentation, you do not want your listeners flipping back and forth to make sure they have not missed something—because then they *will* miss something! If you have complex information to communicate, duplicate it for your listeners, tell them during your presentation what it is that you have given them, and encourage them to read it later.
- *A presentation is also a visual medium.* This is so in a number of respects, but at this point we are talking about your visual aids, which will probably be slides or one sort or another projected on a suitable screen or wall. Here are two things to remember:

  - Every slide must be easily readable from everywhere in the room. Don't even risk it—go to the presentation room and run through your slides, viewing them from the farthest corner. If you can't read them easily, your listeners probably won't bother to read them at all.
  - Graphics such as charts, graphs, or equations should be understandable at a glance. Under no circumstances should it take the listener more than two or three seconds to grasp the main information contained in a visual. Once again, if you feel you need complicated graphics, you probably have a topic that is too complex for an oral presentation.

- *A presentation is more than a handout!* The relationship among the three common elements of a presentation (oral presentation, visual aids, and handouts) can vary depending upon the topic, the audience, and the occasion. However, if you want your audience to feel that it was worthwhile listening to you, consider these things:

- Your visuals should not just duplicate your handout. Your visual aids may duplicate a portion of your handout, or vice-versa, but simply projecting the pages of your handout onto a screen leaves listeners asking "why couldn't I have read this in my office (in half the time)?"
- Do not simply read your visuals back to the audience. Instead of leaving your listeners with the feeling that the brief, bulletized material on the screen is all you have in your head, show them that your visuals represent just the tip of an impressive iceberg (your mind).

A Presentation as Live Theatre

The principal thing that differentiates a presentation from every other form of business communication is that a presentation is done "live." It is really a form of live theatre, conducted in a physical space with real people present. Just as theatre people control the environment in which their plays are presented, so you, as a presenter, should exercise as much control as possible over the environment in which you will be presenting. If there are parameters of the environment that you cannot control, then you must design your presentation with these constraints in mind, at the very least.

In short, as you prepare to make your presentation, you should think of yourself as a theatrical producer. Here are some of the elements that you should consider:

- *Sound.* Many things can affect sound projection during a presentation, including the strength of your voice, the size and structure of the space, the carpeting and furniture, and the number of people present. If at all possible, go to the presentation room and do a sound check. Can you be heard easily in the last row? Will you need a sound system? Is one available? Also, remember that a room is more "live" when it is empty than when it is full of people. The point is that you want *every* word to be heard by *every* listener.
- *Physical intimacy.* How close will your audience be when you are speaking? Your demeanor may have to be a little less formal if your listeners are very close.

- *Lighting.* This is often the most problematic element of presentations. You want to leave the lights on, since you are presenting "live," yet you also want your listeners to be able to see your visual aids. By all means go to the presentation space beforehand and make sure that you can adjust the lights so that your slides are easily readable from everywhere in the room.
- *Technology.* If your presentation depends upon an overhead projector, either manual or computerized, plus a screen, and perhaps a PC, do a trial run early enough so that you can obtain substitute equipment (or bulbs, etc.) to replace anything that does not work. Also, are you going to be your own "stage crew," or will you need someone to help you set up and run the equipment? If the latter, make arrangements enough in advance so that there will be no surprises.
- *Staging.* You may not be able to influence this, but consider it anyway. In planning your staging you want to use the room or space as effectively as possible. Is there a raised area or stage that you can use? Are there windows that would cause the sun to be in your viewers eyes if you stood in front of them? Is there a seating arrangement that would be best (or worst) for your purposes? Are there partitions that can be used to change the size of the space to your advantage? Go to the presentation room the day before you are scheduled to present, sit for a while, walk around, and try to think like a stage manager!
- *Schedule.* As with staging, you may not have control over when you present. However, give it some thought. If you are one of numerous presenters, you may wish to request the first or last spot on the agenda. Information presented first and last tends to be remembered more than information presented in the middle. Also, presenting while people are eating is less desirable than presenting either before or after the meal.

Presentation Delivery

Although this is a book about writing, what you have written will not be effective if it is not delivered well. Here are some guidelines for achieving good presentation delivery:

- *Learn to use your voice effectively.* Whether you have a "big" voice or a small one, you can still learn to use it properly and successfully. If you do many presentations, you might consider consulting with a vocal coach. In any case, here are some guidelines for using your voice correctly:

  - *Think of speech as a physical activity.* Just as there is an appropriate "position" for receiving a serve in tennis or lining up in football, so there is a proper physical stance for good speaking. Stand with your back straight and your shoulders back but relaxed. Support your breath from the navel, not from the upper chest. Your facial muscles and jaw should be relaxed, and your throat should be open. To accomplish the latter, place the tip of your tongue against the back of your lower teeth.
  - *Practice voice placement.* Vocal coaches talk about using "the mask," by which they mean the front of the face. Another way of saying this is to talk about "voice placement." Think of your voice as a stereo system. The vocal apparatus in your throat is an oscillator, but the sound it makes must be amplified. This is done by the cavities in your head and the bone structure of your face. To place your voice properly, think of your face as the loudspeaker of your vocal stereo system.

- *Use your body effectively.* Remember that giving a presentation is somewhat like theatre. If you have never taken acting lessons, you might consider it. The point is to be as dynamic a physical presence as you can in front of your audience. Here are just a few guidelines:

  - *Don't turn your back to your audience unless absolutely necessary.* If you need to point to a visual, arrange things so that you can "cheat" a little, achieving your demonstration without completely turning away from your audience. Often, a light pointer can accomplish this for you.
  - *Use your hands . . . but consciously.* Practiced hand gestures that are well timed and appropriate can raise your presentation to another level. Unplanned, repetitive, or meaningless gestures, on the other hand, can distract from your presentation and may even convey insincerity.

- *Don't wander . . . but don't stand still.* Use your space as a stage. In theatre, the planning of actors' movements around the stage is called "blocking." Be your own director and "block" your movements during your presentation. Like hand gestures, a few good, practiced, meaningful movements will add physical magnetism to your presentation and will convey a sense of self-confidence.

- *Use eye contact effectively.* This is not as easy a concept to grasp as it sometimes seems (or as it is sometimes said to be). For one thing, eye contact is somewhat culturally determined. If you are making a presentation to an audience that includes people from another cultural background, find out, if you can, how they feel about things like physical closeness, eye contact, etc. For an American audience, fairly direct eye contact is entirely acceptable, as long as it is not construed as "staring." To strike the right balance and achieve a soft or sensitive style of eye contact with your audience *think of yourself as wearing a hard hat with a miner's lamp on the front.* When you look at someone in your audience, the light shines on them automatically, but the light is not exactly the same as your eye contact. In this way, you give people the pleasure of feeling that you notice them as individuals while not staring directly into their eyes in an offensive way. Another note about eye contact: learn to move your eyes smoothly from one listener to another, without darting your head back and forth. Darting movements imply insincerity and lack of self-confidence—just the opposite of what you want to convey.

- *Control the pace of your delivery.* In music, pace is called "tempo," and the tempo can change many times during a piece. Just so, to achieve a good delivery you should consider some things related to pacing:

  - *Don't rush!* There are a few people who speak too slowly in front of an audience, but not many. The most common mistake is to go too fast. Remember the linear nature of a presentation—the listeners have only one chance to grasp your meaning as you speak.

- *Vary your pacing.* Expert public speakers understand that just as good writing employs sentences of different lengths so good speaking involves vocal variety, including changes in pace. For example, slowing down or speeding up can add emphasis to something that you particularly want your audience to remember. Practice using pacing as one of the tools in your presentation toolbox.

Using Humor

At the beginning of this section we stated that a presentation is a formal type of communication. That does not rule out the use of humor, however. Beginning and ending with a joke or a chuckle can be an extremely effective way of relating to an audience. There are only two caveats when using humor:

- *If you tell a joke, make sure that it is funny.* There is no *requirement* that you be a stand-up comedian. If you can deliver a joke comfortably and get the laugh, by all means use humor. However, if humor is not your strong suit, go for warmth and sincerity.
- *Keep humor to a minimum.* The idea here is that the humor should never get in the way of your content or argument. If it does, save the jokes for the lunchroom.

Do not think that the above caveats are meant as discouragements to the use of humor in oral presentations. The goal is to communicate effectively, and humor is a tried-and-true method of establishing rapport with an audience. For that reason, it is worth trying, if you can manage it. If you feel that you are not the comic type, who can come up with funny things to say spontaneously, script your jokes and practice delivering them. You may even want to videotape yourself, so that you can watch and listen to your delivery. The benefits of humor, in appropriate doses and delivered well, are worth some effort and rehearsal. And remember that nothing charms an audience more than humor that is at least partially directed at one's self! There is no better way in the world to show that you are both comfortable in front of an audience and comfortable with yourself, including both your weaknesses and strengths, than the use of self-deprecating humor.

Memory Solidifier

Instructions: Under each of the headings below, paraphrase (i.e., write in your own words) the main points under that heading in the material you have just studied. If necessary, return to the previous material to refresh your memory.

List six principles of the writing process as they apply to presentations:\

-
-
-
-
-
-

List two things that practicing your delivery will do for you:

-
-

What type of tone is appropriate for presentations?

-

Brevity as a requirement for presentation writing should be considered from what two perspectives?

-
-

Why is the presentation not well suited to communicating detailed or complex information?

-

List two things to remember about using visuals in presentations:

-
-

List two things to remember about the relationship between speech, visuals, and handouts in presentations:

-
-

List six things to consider in order to "produce" a good presentation:

-
-
-
-
-
-

List to important ideas related to the proper use of the voice:

-
-

What three things should you remember about using your body during a presentation?

-
-
-

What mental exercise can help you develop a sensitive but effective style of eye contact?

-

List two points to remember with regard to pacing:

-
-

What two things should you remember about the use of humor?

-
-

Appendix

Writing Tune-ups

Writing Tune-up No. 1

Using Hyphens

Someone once wrote that "anyone who takes the hyphen seriously will surely go mad." It's true! But there are some principles that we can use to make those "head-scratcher" decisions.

Actually, the hyphen causes most of its problems in two specific areas: compound modifiers and prefixes. Let's look at these two cases.

A compound modifier consists of two or more words that together modify a noun. Here's an example: "the state-of-the-art product." When the compound modifier comes *before* the word it modifies, it is usually hyphenated. However, in many compound modifiers the first word is an adverb, and many adverbs are formed by adding "ly" to an adjective (e.g., "quick" and "quickly"). If a compound modifier comes before the word it modifies but the first word ends in "ly," do not hyphenate (e.g., "the perfectly engineered component"). Now for the curve ball. When a compound modifier comes *after* the word it modifies (i.e., after a "state-of-being" verb such as "is" or "was"), the hyphen is also not used (e.g., "the house was well built").

The other problem with the hyphen concerns prefixes such as "multi," "anti," "non," etc. The general rule is to run the prefix together with the main word, but the *Chicago Manual of Style* also lists several cases in which a hyphen should be used to set off the prefix: (1) when the main word begins with a capital letter or a numeral (e.g., "non-Treasury bonds," "pre-1914"); (2) when there is a doubling of a consonant or vowel (e.g., "non-native," "meta-analysis"); and (3) when a prefix stands alone (e.g., "over- or underused"). My own suggestion is that you also consider *readability*. With prefixes attached to long, complicated words, it may be better to hyphenate just so the reader can visually grasp the term more easily. Take this test: Which is easier to read at a glance—"multiconfigurational" or "multi-configurational"? I'll let you decide—just remember to be consistent within your document.

Writing Tune-up No. 2

The Ubiquitous Comma

All of the technical aspects of language contribute to communication—vocabulary (including spelling!), sentence structure, and, yes, punctuation. In fact, my vote for one of the most important elements of English in terms of clear communication is the "lowly" comma. I have tried to elevate its importance in my title by calling it "ubiquitous" because it has many uses and appears almost everywhere in written English.

One way to think of the comma is in comparison with the period. If the period is a red light, or "stop" sign, the comma is a yellow light or "yield" sign, which we use to tell the reader to pause and consider the next sentence element in relation to what has come before.

Let's think for a moment about the origin of language. It began with people trying to communicate their ideas and feelings to others through (initially) verbal and (later) written symbols. We won't go back to the very beginning. Instead, let's start at the point where the basic written sentence had emerged as an encapsulation of a single thought—one independent clause with a subject and a verb . . . or what we call today a simple sentence. Here is an example:

"They went camping."

Lots of things about life can be described and communicated with simple sentences. However, before long people must have wanted to communicate more complicated messages. For example, two independent thoughts might be related in some way, and so they might be encapsulated in what we call today a compound sentence: two independent (stand-alone) clauses connected by a word that indicates how they are related. Today, we generally form compound sentences by using one of the coordinating conjunctions (and, but, for, so, or, nor, yet)—preceded by a comma:

"They feared it would rain, but the weather was good."

Two independent but related thoughts are now joined by a comma and the conjunction "but."

A complication that I discuss in my article on sentence structure is the situation in which there is only one subject but more than one verb following it. Here are two examples to show the difference—and note the comma usage:

"They camped beside a stream, and they enjoyed a relaxing day." (two independent clauses, each with a subject and verb, connected by a comma and the conjunction "and"—i.e., a compound sentence)

"They camped beside a stream and enjoyed a relaxing day." (one subject—"They"—followed by two verbs—"camped" and "enjoyed"—which is a simple sentence with a compound predicate, requiring no comma)

Probably later in language evolution, people began to construct sentences that contained at least one independent clause (contains a subject and a verb and can stand alone as a complete thought) and one or more "dependent" or "subordinate" clauses (they contain a subject and a verb but cannot stand alone as an independent thought). These dependent clauses are generally connected to the main clause by a subordinating conjunction such as "because," but in analyzing these types of sentences we need to consider the subordinating conjunction as part of the dependent clause, not just as a connector. Here's the general rule: if the subordinate (dependent) clause comes before the main (stand-alone) clause, it is set off with a comma, but if it comes after the main clause it does not normally need to be set off with a comma. Here are two examples:

> "They hiked in the woods because they wanted to get some exercise."

"Because they wanted to get some exercise, they hiked in the woods."

These are what we call "complex" sentences—one or more independent clauses plus one or more dependent clauses.

(For an important situation that involves a special type of subordinate clauses, see my article on "relative clauses," which tells you how to punctuate those types of sentences.)

Another frequent use for commas is in setting off introductory material in sentences. This can be a word, a prepositional phrase, or, as we saw above, a dependent (subordinate) clause. Here are some examples:

"However, it rained the next day." (introductory word)

"During the storm, they played cards in their camper." (introductory prepositional phrase)

"Because it rained, they could not take another hike." (introductory dependent clause)

The third common usage of the comma is to separate the elements of a series. Interestingly, series elements can be clauses or phrases as well as words. However, perhaps the most familiar series elements consist of a string of adjectives that modify a noun. Example:

"After the rain, they enjoyed the clear, fresh, and cool air." (the adjectives modify "air")

Note that I used a comma after "fresh." In general usage, it is not necessary to put a comma between the last two elements of a series unless there is a possibility of misreading. However, in scholarly and technical material I always suggest putting in this "serial" comma to assure that there is no confusion. In fact, why not put it in all the time—it's not ungrammatical, and it guarantees that you will not be misread.

There are other uses of the comma, but if you master these three you will be right most of the time. You might also want to read the article on sentence structure for more specific information about punctuation of clauses.

Writing Tune-up No. 3

Sentence Structure Problems

As a reader yourself, you certainly know that confusing sentences can foul up communication. Let's review the basic principles of sentence structure in English.

Take a deep breath. I think the best way to get a handle on sentence structure is to jump right into the fray—complex sentences. Memory jogger: a complex sentence has at least one *independent* clause (it has a subject and a verb and can stand alone as a complete statement) and at least one *dependent* or *subordinate* clause (which also has a subject and a verb but doesn't make a complete statement—it can't stand alone). Here are some examples:

> "Tom enjoys playing basketball." (independent clause, as well as a complete statement/sentence)

> "Because he is tall . . ." (has a subject and a verb—"he" and "is"—but doesn't make a complete statement.

As we put these two examples together into one sentence, we need to consider one of the main problems with complex sentences—punctuation.

Rule #1: If the dependent clause introduces the sentence, it is set off with a comma. Example:

> "Because he is tall, Tom enjoys playing basketball."

Rule #2: If the dependent clause follows the independent clause (often called the "main" clause), it is generally *not* set off with a comma. Example:

"Tom enjoys playing basketball because he is tall."

(For an exception to this rule, see my Writing Tune-Up article on relative clauses.)

At this point, you're almost an expert on sentence structure. Now, here's another type of sentence that you use a lot—compound sentences. A compound sentence contains at least *two* independent clauses, which are normally connected by one of the *coordinating conjunctions* (and, but, so, for, or, nor, yet) preceded by a comma. They can also be connected by a conjunctive adverb such as "however" (preceded by a semicolon and normally followed by a comma) or even by just a semicolon.

Examples:

"Tom strained a muscle, so he could not play in the basketball finals." ("so" is a coordinating conjunction that connects the two independent clauses)

"Tom had wanted to play in the finals; however, his strained muscle prevented it." ("however" is a conjunctive adverb that also connects the two independent clauses)

"Tom had wanted to play in the finals; his strained muscle prevented it." (Use this type of construction *very* sparingly because it makes comprehension more difficult.)

This seems pretty straightforward, but one of the trickiest sentence structure problems occurs when a sentence has more than one verb but only one subject.

"Tom can play defense and can shoot well."

Notice that there is no comma. That is because this is really a *simple* sentence—one independent clause with two verbs that both have the same subject (Tom). To see the difference between this example and a compound sentence, we just need to add a second subject:

> "Tom can play defense, and *he* can shoot well."

Because of the second subject, we have a compound sentence, which requires a comma.

Then, there are compound-complex sentences, which consist of a compound sentence (two or more independent clauses) with one or more dependent/subordinate clauses added. Here's an example:

> "Because he is quick on his feet, Tom is good on defense, but he can shoot well, too." (subordinate/dependent clause . . . independent clause . . . independent clause)

Luckily, all of the principles outlined above also apply to compound-complex sentences. And there's even an added benefit—putting in a few compound-complex sentences can make your writing more interesting to the reader and more powerful. So, go for it!

Writing Tune-up No. 4

Relative Clauses

Relative clauses give us all heartburn in English. To begin, let's answer the question: What IS a relative clause? In a nutshell, relative clauses are a subset of subordinate clauses (i.e., they have a subject and a verb but cannot stand alone) that are introduced by one of the five relative pronouns: that, which, who, whose, or whom. They most often fall at the end of a sentence or (and this is the strange part) right in the middle of a main or independent clause. Here are two examples:

"George is the man *who lives on the corner*."

"Bill, *who is tall*, likes to play basketball."

I'm sure you don't have difficulty constructing sentences like these, but the problem is punctuating them. As you can see, in the first example the relative clause was not set off with a comma, while the relative clause in the second example was. The reason is not their location in the sentence but rather the roles they play in the sentences.

The distinction we need to make is between "restrictive" and "nonrestrictive" relative clauses. In the first sentence, the clause "who lives on the corner" is restrictive, or essential, to the sentence. In other words, if it were taken out, the main clause would lose some significant meaning. In the second sentence, the relative clause is nonrestrictive, or nonessential. If you take out the clause, the main clause still conveys the same meaning.

Now, what about the punctuation? Just remember that restrictive or essential clauses ARE NOT set off with commas, while nonrestrictive or nonessential clauses are set off. An easy way to get this right is to see if the relative clause sounds like it could be put in parenthesis. If it does, it's probably nonessential and should be set off. Examples:

"The paint (which had just been applied) was not yet dry.

"The paint, which had just been applied, was not yet dry.

The other problem with relative clauses involves the use of "that" and "which." It's easy enough if you just remember that "that" introduces restrictive clauses (no commas), and "which" is used to introduce nonrestrictive clauses. Here are two examples.

"The car *that is in my neighbor's driveway* looks new." (restrictive)

"Tom got a speeding ticket on the way to work, *which is why he was late.*" (nonrestrictive)

The final thing to understand is why the five introductory words mentioned above are called relative "pronouns" rather than "conjunctions," which normally connect clauses. The reason is that these words often wear two hats—as a connector and as the subject of the clause (which can only be a noun or pronoun). For examples, return to the first two sentences above, in which "who" serves not only as a conjunction/connector but also as the subject of the clauses.

Writing Tune-up No. 5

Nonsexist Language Solutions

In the past (prior to about 1970), when we referred to "people in general" in English, we normally used masculine pronouns (e.g., "Any employee wanting to change *his* medical plan should go to the Human Resources Department."). However, as you are undoubtedly aware, we now try to use "nonsexist" language in such situations. There are three good ways to deal with this issue.

The first method is to use constructions such as "his or hers," "he or she," etc. Admittedly, these are somewhat cumbersome, especially if they need to be repeated numerous times in a document. For that reason, some new constructions have emerged such as "he/she" and "s/he." My sense is that these constructions have not yet been accepted as formal English, but they may very well be in the future, and I would suggest that you watch the journal articles you read for signs of this. For the present, if you have just a few sexist-language problems in your document, I recommend using "he or she," etc. If it starts to sound redundant, try one or both of the following techniques.

- Pluralize the sentence so that you can use "gender-neutral" plural pronouns (e.g., All employees wanting to change *their* health plans . . .).
- Alternate using "him," and "her," etc. during your document. This conveys to the reader that you are aware of the issue of sexist language but do not want to sound redundant.

In addition, a client of mine who is a business scholar recently came up with an elegant solution that may be applicable in some cases. He stated up front that "buyers" in his paper would be referred to with feminine pronouns, while "sellers" would be considered male. Problem solved.

There is no going back on the issue of nonsexist language, yet we are all feeling our way along as time passes. I'd love to hear about any new solutions you may encounter.

Writing Tune-up No. 6

Verb Tense When Referring to Past Research

There is a tradition in academia of referring to past research with present tense verb forms—e.g., "Jordan (2001) states that . . ." rather than "Jordan (2001) stated that . . . ." I see no problem with this, except that it sometimes puts one in a position of referring to very old sources (though perhaps seminal) in the present tense.

Then there is the problem of referring to one's own current research (i.e., the work the reader is actually reading). For example, should you say, "To test this hypothesis, we perform an analysis . . . ?" Or should it be " . . . we performed an analysis"?

Unfortunately, the style guides are not much help. The *Chicago Manual of Style* gives the following examples:

> "Various investigators (Jones and Carter 1980) have reported findings . . . ."

> "Jones and Carter (1980) report findings . . . ."

The main problem I see in the papers I edit is inconsistency in verb form usage. I have seen no guidelines from academic journals on these issues, so I can't help but feel that both past tense and present tense forms are acceptable. My only suggestion would be to make conscious decisions in these matters and then try to stick with them as much as possible, especially in places (such as the literature review section) in which multiple sources are referred to in succeeding sentences or in close proximity.

Personally, I think that past tense makes the most sense in today's language environment, even when referring to your own current paper. However, as with many editorial decisions, I have learned to be flexible. As always, my guiding principles are readability and comprehension. Style consistency can contribute to these things.

Index

A

Adjectives 38
  Adjective placement 38
  Adjectives with acronyms 39
  Articles 39
  comparative form (comparison) 39
  Degrees of comparison 39
  Demonstrative adjectives 39
  Hyphenation 40
  Indefinite adjectives 39
  Interrogative adjectives 39
  Limiting adjectives 39
  Multiple adjectives 40
  Number adjectives 39
  Omission of articles 40
  Positive form (comparison) 39
  Possessive adjectives 39
  Relative adjectives 39
  Superlative form 39
  Unit modifiers 40
  Usage 39
Adverbs 44
  Adverb placement 45
  Adverbials 45
  Common adverbs 45
  Conjunctive adverbs 45
  Interrogative adverbs 45
  Numeral adverbs 45
  Split infinitives 46
  Squinting adverb 45
  Types of adverbs 45
  Usage 45
Audience 109
Alphanumeric format 103

B

Basics of English 13
Brevity 131
Business letter, formal 146
  Letter format 146
  Tone 146
Business writing 115
  Identifying the audience 117
  Importance of brefity 131
Business writing "tone" 122
  Conveying negative information 126
  Defining tone 122
  Expressing enthusiasm 125
  Formal tone 122

Informal tone 124

C

Capitalization 88
   Capitalize sparingly 88
   General rules 88
   Hardware names 88
Cautions 106
Conjunctions 55
   Conjuctive adverbs 57
   Coordinating conjunctions 55
   Correlative conjuctions 56
   Standard coordinating conjunctions 55
   Subordinating conjunctions 57

D

Development, methods of 140

E

E-Mail Correspondence 134

F

Fonts and typefaces 104
Format 102
   Alphanumeric format 103
   Figure Format 105
   Justification 104
   Line lengths and margins 104
   Military decimal format 102
   Standard formats 102
   Table format 105
   Typefaces and fonts 104
   Warnings, cautions, and notes 106

G

Grammar and punctuation 84
   Capitalization 88
   Punctuation, technical writing 89
   Syntax and usage 84

I

Identifying the audience 117
Internal report 156
Interoffice memorandums 136
   Memo content development 139
   Memo format 137
   Methods of development 140
   Outlining 142
   Readability issues 138

L

Letter content 150
   Closing paragraph 151
   Introduction 150
   Substantive paragraphs 151
Letter format
   Additional information 149
   Body of letter 148
   Closing 148
   Contact information 146
   Date of correspondence 146
   Greeting 147
   Internal address 147
   Pagination 149

M

Main idea in main clause 67
Methods of development 140
   Cause and effect 141

Chronological 140
Comparison 141
General-to-specific 141
Order of significance 141
Sequential 140
Spatial 141
Specific-to-general 141
Military decimal format 102

N

Nomenclature 94
　Capitalization 94
　Clarity and comprehension 94
　Controls and indicators 94
　Lights and LEDs 95
　Pots and selector switches 95
　Readouts and meters 95
　Switches 94
　Verify that 95
Notes 106
Nouns 15
　Abstract nouns 15
　Appositive 16
　Collective nouns 15
　Common nouns 15
　Compound nouns 17, 18
　Concrete nouns 15
　Forming plurals 17
　Gerunds 16
　Linguistic functions 16
　Mass nouns 15
　Noun usage 17
　Object of a preposition 16
　Objective complement 16
　Objecto of a verb 16
　Possessive forms 17
　Proper nouns 15
　Subject of a sentence 16
　Subjective complement 16
　Types of nouns 15

O

Outlining 142

P

Parallel construction 68
Parts of speech 15
　Nouns 15
Phrases 65
　Noun phrases 65
　Prepositional phrases 65
　Verb phrases 65
Prepositions 50
　Idiomatic expressions 52
　In titles 52
　Linguistic functions 50
　Prepositional phrase 50
　Usage problems 51
Presentations 176
　As "writing" 176
　As live theatre 179
　Characteristics 177
　Delivery 180
Pronouns 21
　Case 24
　Demonstrative pronouns 21
　Gender 25
　Indefinite pronouns 22
　Intensive pronouns 23
　Interrogative pronouns 22
　Number 25
　Personal pronouns 21
　Pronouns with gerunds 23
　Reciprocal pronouns 23
　Reflexive pronouns 23

Relative pronouns 21
Types of pronouns 21
Proposal content
  Cost 169
  Technical solution 168
Proposals, external 166
  Boiler plate material 171
  Editing 173
  Graphic design 173
  Proposal content 167
  Proposal writing as "writing" 172
Punctuation in technical writing 89
  "Em" dash 90
  "En" dash 90
  Hyphen 89
  Minus sign 90
  Quotation marks 91
  Serial comma 91

R

Readability 138
Report structure
  Abbreviations and acronyms 160
  Abstract 157
  Appendixes 161
  Body 159
  Conclusions and recommendations 160
  Introduction 158
  References 161
  Results 160
  Summary 158
Report, internal
  Report structure 157
  Report style 156
Revising 112

S

Sentence fragments 70
Sentence structure 62
Short forms 80
  Articles with acronyms 80
  Degree symbol 81
  English and metric units 81
  First use 80
  Initial caps 80
  Period with units of measure 82
  Plurals 81
  Standards 82
  Units of measure 81
Syntax and usage 84
  "In/Into, on/onto" 87
  "Which/That" 86
  Comprised of/comprises 84
  Data 84
  Interface 84
  Mathematical operators 84
  Noun clusters 85
  Numbers beginning sentences 85
  Prefixes 86
  Spelling Out Numbers 85

T

Table format 105
Technical writing 75
  First two laws 79
  Fourth law 111
  Third law 109
Technical writing style 97
  "It" as a subject 98
  Active voice 97
  And/or 97
  Clarity and conciseness 97

Flammable/inflammable 98
Introductory clause 98
One action per step 98
Procedural steps 98
Specific verbs 99
Starting sentences 99
State actions specifically 99
Telegraphic style 99
Typefaces and fonts 104
Types of sentences
  Complex sentences 63
  Compound-complex sentences 64
  Compound sentences 63
  Simple sentences 63
Types of Sentences 62

U

Use simple sentence structure 67

V

Verb phrases
  Gerund phrases 66
  Infinitive phrases 66
  Participial phrases 66
Verbs 28
  Active voice 33
  Auxiliary verbs 32
  Characteristics of verbs 30
  Finite verbs 29
  Forms of verbs 29
  Imperative mood 34
  Indicative mood 34
  Infinitive 29
  Intransitive verbs 28
  Mood 34
  Nonfinite verbs 29

Number 30
Participles 30
Passive voice 33
Person 30
Subjunctive mood 34
Tense 32
Transitive Verbs 28
Types of verbs 28
Voice 33

W

Warnings 106
Writing process
  Audience 109
  Creating 112
  Organizing. *See*
  Organizing 111
  Reading 112
  Revising 112
  Writing as a process 111